Wood Characteristics

Christoph Richter

Wood Characteristics

Description, Causes, Prevention, Impact on Use and Technological Adaptation

Springer

Dr. Christoph Richter
Opitzer Weg 20
D-01737
Tharandt
Germany

Based on an updated translation of the original German 3rd edition of "Holzmerkmale" published by DRW-Verlag Weinbrenner, Leinfelden-Echterdingen (2010)

ISBN 978-3-319-07421-4 ISBN 978-3-319-07422-1 (eBook)
DOI 10.1007/978-3-319-07422-1
Springer Cham Heidelberg New York Dordrecht London

Library of Congress Control Number: 2014955154

Printed on acid-free paper

Springer is part of Springer Science+Business Media (www.springer.com)

Preface

It was 1995, in Dresden, Germany. I had stopped to linger over the display in a bookstore window. Suddenly, my eye caught the cover of a book with two pictures of spruce discs. My curiosity as a forester was piqued. The title read, "Holzfehler – Die Abweichungen von der normalen Beschaffenheit des Holzes" ("*Wood Defects – Deviations from Standard Wood Structure*"), by Hermann Knuchel. I hesitated, because, as far as I knew, the book had been published last century, back in the 1930s. This had to be a new version of the 1934 edition. But why a reprint after 60 years of advances in wood research? Was there no more recent work of its kind?

Investigating further I learned that since Knuchel's work, many studies had indeed touched on the topic of "wood defects" with some surprising new findings, but none provided such a systematic overview as Knuchel had in his day. "Wood defects" were either the subject of in-depth scientific research, carried out by specialists and published for specialists in scientific journals, or they appeared as subtopics in broader works on wood and wood processing, never as the main subject. Yet, the impact of "wood defects" on price and intended use plays just as important a role in the marketing and manufacturing of wood today as they did back in 1934.

Since that memorable day in front of the bookstore window, I've been nurturing a dream to publish a revised study on "wood defects"; at the same time fully aware of the difficult balancing act between covering the topic in sufficient scope and breadth while maintaining the desired technical and scientific depth.

I began by reflecting on the term "wood defect" and the unfortunate way it can stigmatize wood. Can't the same characteristic that prevents the wood from being used for a specific purpose actually make it suitable for another? Of course. For this reason, going forward I began using the more neutral term "wood characteristics".

The book begins by discussing the "General factors leading to the formation of wood characteristics". These influences are responsible for the diversity among the wood characteristics.

The individual characteristics are then categorized into four groups of wood characteristics.

1. *Wood characteristics inherent in a tree's natural growth.*
   These include changes to a tree's stem contour, limbiness and anatomical structure.
2. *Biotically induced wood characteristics.*
   Involving all tree internal and external wood characteristics created by micro-organisms, animals and humans or plants.
3. *Abiotic induced wood characteristics.*
   Wood characteristics created by heat, cold, humidity, wind and other external forces.
4. *Crack forms and causes*, where different causes can lead to cracks with similar forms or different forms can have the same causes, are assigned to a separate group of characteristics.

The chapters on the individual characteristics generally cover these five questions:

1. How can the characteristic be described (anamnesis)?
2. What are its causes (diagnosis)?
3. How can characteristics be influenced as the tree grows (prophylaxis)?

4. How does a characteristic effect the various ways the wood is used (impact assessment)?
5. How can technology respond to wood characteristics (treatment)?

The discussion on the individual wood characteristics is supported by corresponding illustrations and a separate section of photographs shows examples of how the characteristics typically appear in nature. The English edition of "Wood Characteristics" maintains the same objective as the 3rd 2010 German edition (Richter 2010).

The book addresses all who work with wood professionally. Foresters, gardeners and arborist want to be able observe a living tree and identify its internal features and the causes of any existing wood characteristics. Based on these findings they can determine how to avoid certain undesirable characteristics, or alternatively how to promote favorable characteristics as the tree and stand grow.

My aim is also to address wood technologists seeking to prevent the impact of adverse wood characteristics on wood processing, or enhance any favorable wood characteristics, as the case may be. Lastly, it gives options for technically adapting, handling and processing wood with specific wood characteristics.

Botanists and dendrologists learn how wood characteristics occur, how they affect living trees and wood products, and how they can be either avoided or encouraged.

New to this English edition is a comparison of wood characteristics found in trees from the boreal, temperate and tropical climate zones. The results show a clear relationship between the effects of sunshine duration, the vertical and horizontal angle of radiation, and crown coverage and the way wood characteristics form.

The influence of wood characteristics on wood quality – compiled in numerous national wood grading standards – is discussed to an extent that clearly shows the connection between wood quality and wood price in the timber industry.

The knowledge gathered in this book is based on the scientific and practical work of foresters, wood technologists and biologists spanning many generations. Without them, but also without the more recent generous support of certain people and institutions, this edition of the book could certainly never have been completed. Therefore I extend my special thanks to Michael Köhl, Institute for World Forestry at the University of Hamburg, for encouraging me to pursue this new edition; Gerald Koch and Hans-Georg Richter (Thünen Institute of Wood Research Hamburg) for supporting me with their wood science expertise.

I would like to thank the German Federal Ministry for Food and Agriculture (BMEL) for providing the material basis for the necessary research in the tropics; the staff at the Centre for Agricultural Research (CELOS), the Stichting voor Bosbeheer en Bostoezicht (SBB) as well as Jos Dennebos, Herman Fräser and Rasdan Jerry (E-Timberindustry) in Surinam, along with my colleagues Bernhard Kenter, Timo Schönfeld and Lars Niemeier (University of Hamburg), who helped make the wood science research in Surinam possible.

My great appreciation is extended to my fellow colleagues from the School of Forestry Management at the Technical University of Dresden, especially Claus-Thomas Bues and Ernst Bäucker, for the photographic material they provided and the insights I gained from them during our numerous professional discussions.

I also sincerely thank Susan J. Ortloff (Oregon, USA) for her sensitive translation. The financial resources for this purpose were mainly provided by the University of Hamburg and the BMEL.

Representative for professional cooperation with Springer-Heidelberg, I thank Christina Eckey (Senior Editor, Plant Sciences) and Anette Lindqvist (Production Coordinator) for edition from "Wood Characteristics".

Last but not least, I thank my wife, Dorothea, for her many years of patient understanding when quite often, instead of spending time with her, I spent it entrenched in this project.

Tharandt, Germany Christoph Richter
Summer 2014

Contents

Part I

Wood, a Truly Remarkable Material

1 The Anatomical Structure of Wood

Wood consists of different types of cells, each with specific functions: transportation, support, and storage. At 20× magnification, it is easy to see how the various cells in a softwood sample together create a distinct structure (Fig. 1.1). The most common cell type, longitudinal tracheids, transports water up the length of the trunk, giving the tree stability. Wood ray tracheids transport nutrients in a radial direction. Longitudinal parenchyma cells store food reserves, while wood ray parenchyma cells support the exchange of material both radially and to neighboring tracheids.

In evolutionarily younger hardwoods, the cells are even more specialized (Fig. 1.2). The vessel cells bind to create highly efficient water pipelines. The narrow-lumined vessel tracheids transport water. The wood fibers mainly provide stability. The longitudinal and pith ray parenchyma cells both transport and store nutrients.

The cells mainly run longitudinally up the length of the tree stem. This leads to anisotropy with differing wood properties in the longitudinal, radial, and tangential directions. This also leads to variations in mechanical stability. As such, the wood's tensile strength is nearly two times higher than its compressive strength. The bending strength, a combination of tensile and compressive strength, lies somewhere in between. There is considerable difference between the strength of the wood running along and against the fibers. Among all tree species, the ratio of tensile strength

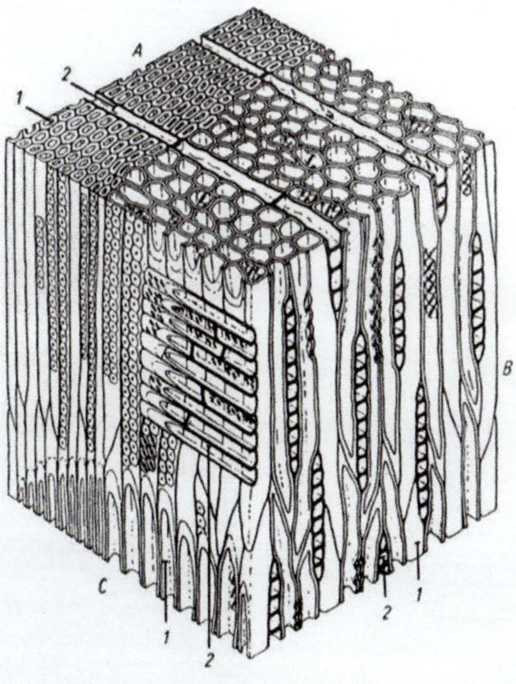

Fig. 1.1 Microscopic structure of softwood. Wagenführ (1966) from Oliva in Tortorelli. *A* Cross section, *1* tracheids, *B* tangential section, *2* wood rays, *C* radial section

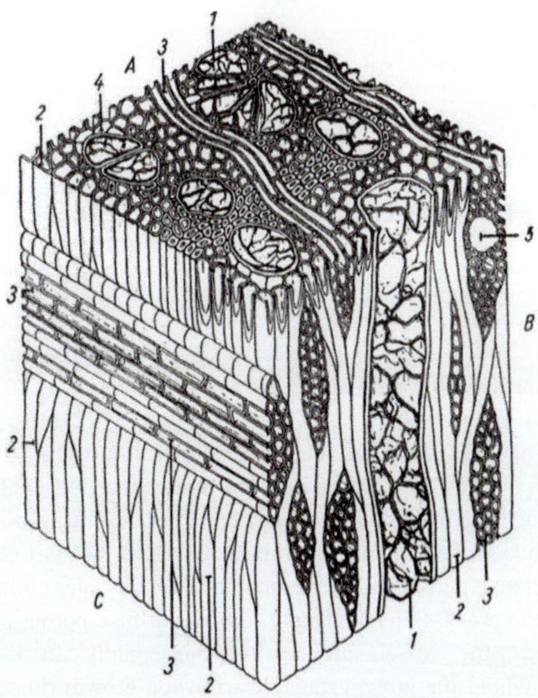

Fig. 1.2 Microscopic structure of hardwood. Wagenführ (1966) from Oliva in Tortorelli. *A* Cross section, *1* vessels with tylosis, *B* tangential section, *2* libriform fibers, *C* radial section, *3* wood rays, *4* longitudinal parenchyma

C. Richter, *Wood Characteristics: Description, Causes, Prevention, Impact on Use and Technological Adaptation*,
DOI 10.1007/978-3-319-07422-1_1, © Springer International Publishing Switzerland 2015

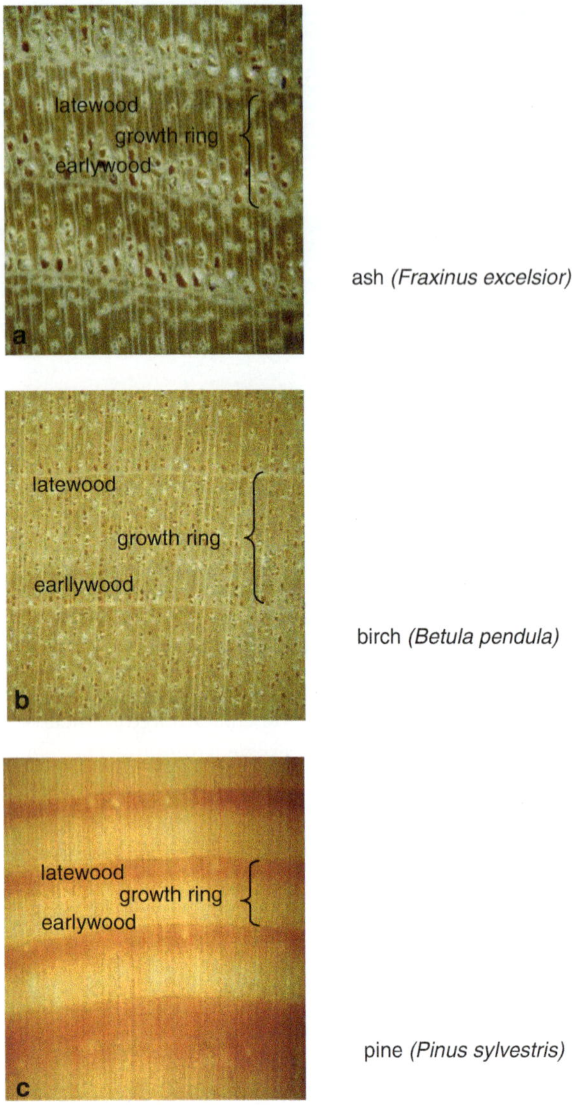

Fig. 1.3 Cross section of a ring porous (**a**, ash (*Fraxinus excelsior hardwood*)), diffuse porous (**b**, birch (*Betula pendula hardwood*)), and a softwood (**c**, pine (*Pinus sylvestris*)). Enlarged: 1:4 (Photos: E. Bäucker)

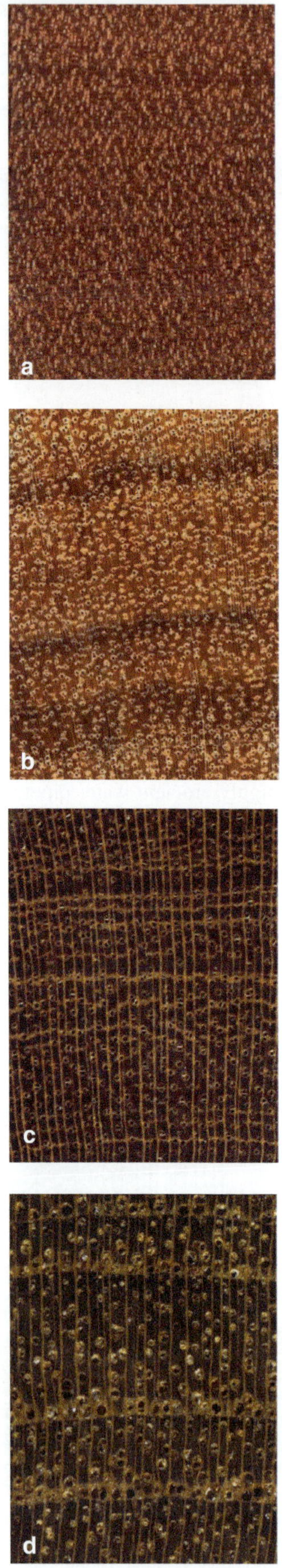

Fig. 1.4 Cross section of tropical woods with variations in increment zone boundaries. Enlarged: 1:10 (Photos: Richter and Oelker (2003). (**a**) Massaranduba (*Manilkara bidentata*): Increment zone visible or only distinguished by density variations in the tissue. (**b**) Cumarú (*Dipteryx odorata*): Increment zone light or indistinguishable boundaries (not to be mistaken for the dark brown stripes). (**c**) Sapeli (*Entandrophragma cylindricum*): Increment zone marked by narrow parenchyma bands. (**d**) Teak (*Tectona grandis*): One of the few ring porous woods in the tropics; increment zone clearly visible by the large pores in the earlywood

along and against the grain is 100:3 … 4, and the ratio of compressive strength along and against the grain is 100:14 … 21 (Niemz 1993).

This anisotropy continues microscopically in the growth rings of trees from temperate regions as new cell layers form during the cambium's vegetative period. Ring-porous hardwood develops wide-lumined vessels at the start of the vegetative period and narrow-lumined vessels later on in the growth process. In much more common diffuse-porous hardwoods, the vessels are smaller, but equally distributed throughout the growth rings. In softwood growth rings, the wide-lumined earlywood tracheids differ abruptly from the narrow-lumined latewood (Fig. 1.3).

In the constantly humid and warm tropical climates, trees typically form increment zones without any significant distinction (Fig. 1.4). Periods of rain or drought, or periods of dormancy or defoliation, appear as increment zone boundaries and are distinguished more or less clearly by variations in cell size, diffused parenchyma bands, and tissue density (Fig. 1.4) (Sachsse 1991). Given the varied growth conditions and the significant variety of species in tropical and subtropical forests, the exact age of tropical trees cannot be determined based on increment zones (Harzmann 1988).

The width and structure of the individual growth ring or increment zone mainly depends on the growth capacity of the specific tree species, nutrient supply, temperature, and precipitation during the vegetation period, as well as on seed years and any damaging events such as drought and insect infestation.

Anyone who has ever chopped wood has taken advantage of this anisotropy. Wood is easiest to cut lengthwise because it splits along the grain. Felled trees split due to the release of internal stress within the wood, starting at the cut and continuing down the length of the stem. Wood rays running in a radial direction also make the wood relatively easy to split from the lateral surface towards the pith. Excellent examples of this are frost cracks on the tree bark. By contrast, wood cannot be split against the grain. The tree's structural building blocks are arranged so that the greatest stability originates in the direction of the trunk's axis.

Wood's significant compression strength and the double as high tensile strength along the fibers enable a tree to hold up under gravity and other external forces. Thus, for example, an ancient spruce (*Picea abies*) growing on a mountaintop has the strength to withstand several tons of snow and ice. Witnessing a tree being hit by a strong gust of wind also offers an excellent example of why a tree needs to have its greatest stability along its fibers.

In tropical primary and secondary forests, heavy crown competition leads to slender, solid wood trees. Forces applied to the foliated crown result in extreme bend and torsion effects, to which the tree reacts by building reaction wood, spiral-grained wood, or uniquely formed stem surfaces.

Only by understanding wood's anatomical structure is it possible to understand why a specific wood characteristic forms and how it affects the way the wood is put to use. Ultimately, every question regarding wood characteristics is actually a matter of wood anatomy. This is true for timber from both temperate zones and the tropics.

Comparing the anisotropic material of wood with other materials clearly shows that the latter have homogeneous microscopic structures (metals, glass) or that the structural elements (chips, fibers in plate materials, and mineral formations in layered rocks such as slate and gneiss) are homogeneously distributed more or less into two levels. This homogeneous structure of amorphous materials, or the layered structure of particle materials or rocks, is often preferred over wood because it is easy to access their material and processing properties. These materials are "predictable."

Nevertheless, clear, defect-free wood expertly used also has excellent performance characteristics. So, for example, the breaking length of wood fibers (length, at which a stick breaks under its own weight) is 15,000 m; in steel St37 with same cross section, it is only 4,700 m (Bosshard 1984a).

Throughout history, the biggest self-supporting vaulted ceilings were made from timber, not concrete. The world record for the heaviest aircraft cargo load was set by the Spruce Goose, a wooden airplane built in the United States to transport troops during the Second World War (Matzek 1985). This record stood for six decades only to be broken by the high-tech Airbus A 380 with a loading capacity of 853 people (Spiegel 2006).

Enthusiasm over wood's truly remarkable properties, however, often fades in practice when a characteristic surfaces making the wood difficult or even impossible to use: A branch within a frame limits the calculated fatigue strength; a batten with missing wood fibers cracks under stress; a stained veneer sheet is unsuitable for high-quality use. Repeatedly, these often unexpected defects threaten to spoil the reputation of wood as a reliable working material. These characteristics hidden inside the wood, or often clearly visible on the stem surface, vary from the tree's "normal" growth or from the "normal" structure of the wood and can significantly influence how the wood is used. Therefore, it is important for anyone working with wood to be familiar with main wood characteristics, how they form, how to prevent them, and how they impact the quality of the end product and the wood's potential technical adaptation.

2 Wood Characteristic or Defect?

The way a tree looks on the surface and its internal features are in fact determined by a set of characteristics. These characteristics form during the natural growth cycle. They include stem shape, branch formation, bark features and the anatomical structure and color of the wood. Throughout the seasons, external events such as temperature variations, rain, snow, wind and lightning also affect the tree. Biotic influences caused by fungal and insect attack, animals, plants, and human activities also play a role.

Surface and internal features make each tree unique. These characteristics are not necessarily defects. Whether they are viewed as (neutral) characteristics or defects is simply a matter of perspective.

Possible approaches are:

From a tree's perspective, a characteristic is only a defect if it significantly influences the tree's natural life expectancy.

Rot can weaken a tree's stability or impair vital functions. A low stem break can kill the trunk with an abrupt loss of crown and foliage (Fig. 2.1). Yet an unusual stem shape, a knot, or obviously the branches, which play an indispensable role in the assimilation process, would not be considered defects.

Fig. 2.1 The fork break in this beech (*Fagus sylvatica*) (*left*) and the deeply imbedded rot in a Gronfolo (*Qualea rosea*) (*right*) in the tropical rain forest are life-threatening defects from the trees' "perspective"

From a woodworker's perspective, the characteristics are defects if they make the wood difficult or impossible to use for a specific purpose (Fig. 2.2).

As a result, a wood characteristic is not a defect if it does not interfere with the wood's intended purpose or if it renders the wood useful for a specific purpose (Fig. 2.3).

Fig. 2.2 Not all branches are equal: This limby spruce will (only) provide lumber full of ingrown and black knots. A branch – at least one with the dead black knot – is seen as a wood defect

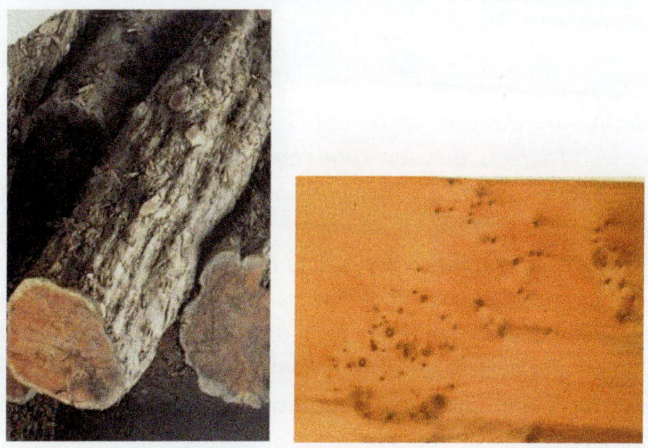

Fig. 2.3 A yew (*Taxus*) stem with many small branches (twigs, suckers) can be used to make valuable burl veneer. In this case, the cluster of knots is seen as a desirable wood characteristic

C. Richter, *Wood Characteristics: Description, Causes, Prevention, Impact on Use and Technological Adaptation*, DOI 10.1007/978-3-319-07422-1_2, © Springer International Publishing Switzerland 2015

Fig. 2.4 After 1,000 years of growth, Germany's oldest oak (*Quercus robur*) is beyond any consideration of use (Ivenack, Germany)

Fig. 2.6 German Spessart oak (*Quercus*) is valued for its straight, clear stem with regular growth rings and flesh-colored wood – excellent for high-quality furniture and cabinet making

Fig. 2.5 In shipbuilding, care was given to use naturally shaped wood for the various parts of a vessel – i.e., branch forks (*red markings*), crooked branches, and curved stems. The photo to the *right* shows frames and stern posts made from an oak tree for the replica of a Viking ship (Roskilde, Denmark)

Happy are the ecologists and aesthetes. Where others see wood defects, they see wood characteristics, special traits, unique to a tree and reflecting a synergy among biozones. They accept them as an expression of nature: diversity of shape, originality, vitality, and passage of time (Fig. 2.4).

The following chapters discuss wood from the viewpoint of the woodworker. For the most part, the neutral term "wood characteristic" is used. Only when a specific feature interferes with an intended purpose will the negative term "defect" be used.

Since time immemorial, people have determined the ideal shapes and properties of a tree stem or a piece of wood based on the ultimate end product. In the Stone Ages, wood used to make a spear had to be straight, slender, and elastic, while wood intended for the handle of a flint ax needed to be solid with a hook shape. Carpenters of the Middle Ages preferred oak for beams because wide growth rings made the wood more resistant to bending. In Lapland, people made sturdy sled runners from sickle-shaped root crowns. And well in to the nineteenth century, tree parts selected for their naturally formed shapes were highly prized in shipbuilding (Fig. 2.5). In modern times, given the constant improvements in manufacturing, solid, straight-stemmed, branch-free trees have become the preferred standard. Demands on stem quality are greatest in furniture and cabinet making (Fig. 2.6).

The wood's specific end purpose, therefore, determines whether a wood characteristic is considered a defect, a minor variation, or even a desired feature. Wood has quality when it is suitable for a specific end purpose. Thus, it is essential that a woodworker has a good understanding of the basic wood characteristics. Some wood characteristics are either directly visible or indirectly apparent on a live tree and therefore are given special attention. This is partly necessary because, on the one hand, early identification saves time and energy spent processing unsuitable wood. And on the other hand, recognizing a desirable wood characteristic early on can result in the wood being graded for a much higher quality product. Timber experts and wood technologists have been searching for effective ways to accurately predict the quality of the processed wood based on the quality of the timber. Basic guidelines, such as the Swiss OPS or the Swiss Timber Industry Standards, rate stem quality in the lower portion of the stem (near 8 m high) in three groups, optimal, satisfactory, and poor, and are capable of identifying 10–30 % of the defects found in the logs (Stepien et al. 1998).

More detailed quality classification procedures currently exist which, while quite time consuming (such as laser scanning), also provide a more accurate quality appraisal for veneer or log grade timber (Schute 1972a, b; Richter 2000; Willmann et al. 2001; Schütt et al. (2005)).

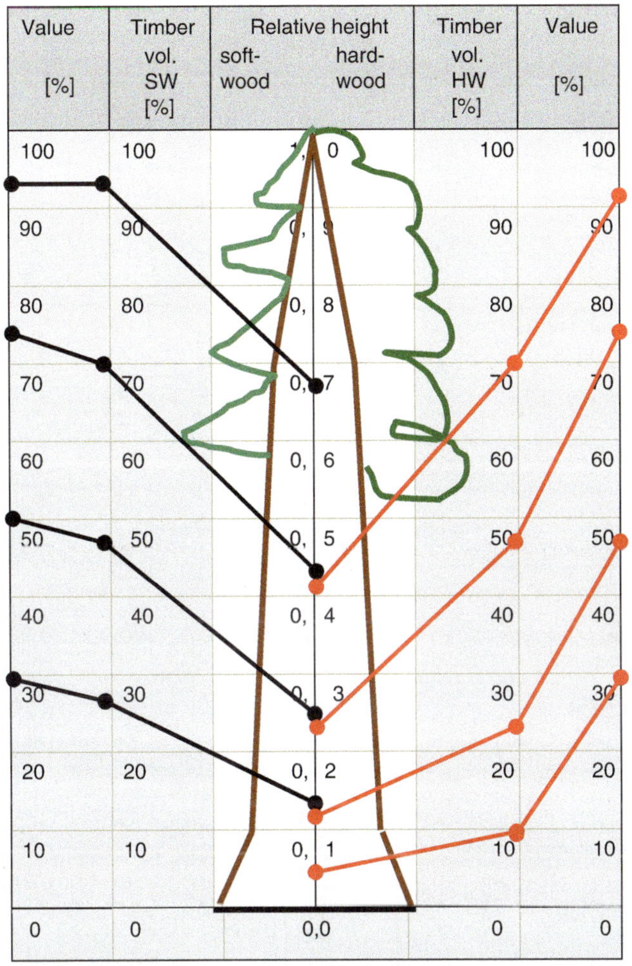

Fig. 2.7 Relationship between volume and value distribution of mature softwood and hardwood trees depending on relative tree height (Richter 2000 after Bachmann 1990)

Stepien et al. (1998) used a multiple regression model to predict the quality of wood from a survey of mature timber with an accuracy of about 60 % for beech, spruce, fir, and pine (*Fagus, Picea, Abies, Pinus*) and about 45 % for larch (*Larix*). The assessment included the ten superficial tree features: branches, suckers, branch scars, bumps, sweep, crook, spiral growth, cracks, necrosis and cancers, (coarse bark only in larch). Only the first 9 m of the stem were assessed in the study, because this section of the stem makes up 50–70 % of the value of softwood and 80–95 % of the value of hardwood, especially beech (Bachmann 1970); compare Fig. 2.7.

Log ends and branch stubs reveal additional, otherwise, hidden characteristics, useful for quality assessment, such as random color variations, decay, pith flecks, growth ring anomalies, reaction wood, and resin ducts. Including as many surface characteristics as possible, the quality of the round timber can be used to predict, with a relatively high degree of accuracy (estimated at around 60–80%), the quality of the future sawn timber or higher-end product. While the cost and expenditure of conducting a quality assessment increases exponentially with the breadth of the survey, failure to conduct a precise quality assessment results in lost revenue. The only way really to mitigate this contradiction is by knowing how to assess wood characteristics.

Timber grading practices vary significantly around the world. The most simplistic method grades logs solely based on length and particularly the average diameter. Today, this method is only used in countries with (alleged) timber surplus and simultaneously low harvest yields. As demand for timber rises, grading standards pay increasingly more attention to features that naturally develop during the course of a tree's life, biotic and abiotic characteristics and crack formations. These grading systems consider characteristics that adversely affect a log's end purpose as defects. Characteristics which allow special usage are considered beneficial. This leads to a differentiated pricing on the international timber market.

In many countries, the requirements for dimension, grade, and end use of commercial timber are set forth in official standards or regulations. Germany followed its own commercial timber grading rules (HKS 2002b) until 2012. The standards recommended for members of the European Union are set forth by the rules on dimension and quality specified by the European Committee for Standardization (CEN) (DIN 1997d, 1998c). There are also bilaterally accepted quality standards established between timber buyers and timber companies that regulate the timber market. As of 2013, German standards follow the framework agreement for sawn timber trade (RVR 2014).

It is impossible to list all the quantitative descriptions given to define individual wood characteristics by the many international grading standards (e.g., Carpenter et al. 1989). Therefore, the following descriptions of specific wood characteristics called or mentioned include extracts from German standards (both past and present), namely, the TGL (TGL 1977b, c), HKS, CEN, and RVR. This seems justified for timber from the temperate latitudes, because Germany's quality standards, established to address a continuously decline in timber supply, date back to the fifteenth century (Willing 1989).

There are three general recommendations for grading tropical timber (Lohmann 2005, p. 87–89):

1. The French grading system, set by the "Association technique Internationale des Bois Tropicaux (ATIBT)," is based on points. Here, a maximum amount of penalty points are assigned to five different quality grades for stems of specific lengths.
2. French classification for "fair merchantable goods" (Loyal et al. Marchande (L & M)). The merchantable timber is classified into five quality grades based on the amount of blemish-free wood, 87.5, 75, 62.5, 100, and 50 %, respectively.
3. English classification according to "fair average quality" (FAQ). The merchantable timber is classified into 5 quality grades based on the amount of blemish-free wood, 100, 90, 80, 70, or 60 %, respectively.

3 General Factors Leading to the Formation of Wood Characteristics

With such variety among wood characteristics, one would think there would also be many different factors leading to their formation. Actually, however, there are only five main "triggers".

The two main "internal" triggers are:

- *Genetic predispositions* and *genetic alterations* in the tree (mutation, genetic defects)
- *Alterations* in the tree's internal *physiological processes* (assimilation, nutrient supply, material transport, chemical reactions, …)

The three main "external" triggers are:

- *Light/radiation* (heliotropism)
- *Mechanical stress* (geotropism, wind, lopsided crown …)
- *Injuries/infections.*

In nature, the effects of these five factors often overlap making a simple cause-effect relationship difficult to identify. Nevertheless, it is important to try to describe how trees mainly react to these five triggering factors before further discussing the individual wood characteristics (formed in response to this "trigger"). These five factors apply, in principle, for trees in the boreal, temperate and tropical zones.

3.1 Genetic Predispositions, Genetic Alterations

All trees grow according to a genetically predetermined design. Hereditary information determines a tree's outward appearance and its internal biochemical processes (Fig. 3.1). If a tree grows under normal site conditions (climate and soil), then woodworkers will usually be satisfied with its morphology. Genetic changes, however, can cause single individuals, or provenances, to deviate from their tree species' "normal form." For example, external characteristics such as the forking tendency among birch (*Betula pendula*) (Fig. 3.2) or the extreme tapering tendency of Engelmann spruce (*Picea engelmannii*) (Fig. 3.3) are genetic. The same applies to fluting in hornbeam (*Carpinus betulus*) (Fig. 3.4) or zwart parelhout (*Aspidosperma excelsum*) (Fig. 3.5), the formation of flanges in elm (*Ulmus laevis*) (Fig. 3.6) and djadidja (*Sclerobium melinonii*) (Fig. 3.7), or burls in spruce (*Picea abies*) (Fig. 3.8) and basralocus (*Dicorynia guianensis*) (Fig. 3.9).

Abrupt changes in a tree's morphology and physiology can also be the result of a *mutation*. A well-known example of a recent, potentially lasting mutation is the corkscrew-

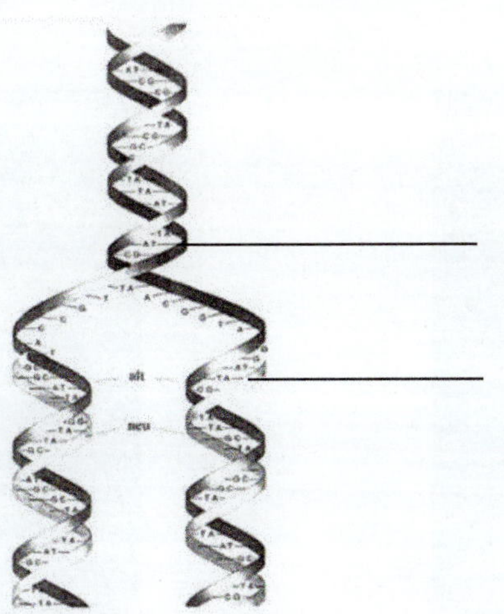

Fig. 3.1 DNA replication (a section of a DNA double helix structure model) (Buchner 2008)

C. Richter, *Wood Characteristics: Description, Causes, Prevention, Impact on Use and Technological Adaptation*, DOI 10.1007/978-3-319-07422-1_3, © Springer International Publishing Switzerland 2015

Fig. 3.2 Genetic tendency to fork in (birch (*Betula pendula*) *left*) or to mono crown (birch (*right*)) (Erz Mountains, Germany)

Fig. 3.3 Genetic tendency to taper in Engelmann spruce (*Picea engelmannii)* (Yukon Territory, Canada)

Fig. 3.4 Fluted hornbeams (*Carpinus betulus*) (Germany)

Fig. 3.5 Fluted witte parelhout (*Aspidosperma marcgrafianum*) (Surinam)

Fig. 3.6 Flanges in a European white elm (*Ulmus laevis)* (Oberlausitz, Germany)

Fig. 3.7 Flanges in djadidja (*Sclerobium melinonii*) (Surinam)

Fig. 3.8 Spruce (*Picea abies*) pimple (hazel growth) (Germany)

Fig. 3.9 Basralocus (*Dicorynia guianensis*) pimple (hazel growth), (Surinam)

Fig. 3.10 Mutation in beech (*Fagus sylvatica, var. tortuosa*) leads to a dwarfed corkscrew shape (Niedersachsen, Germany)

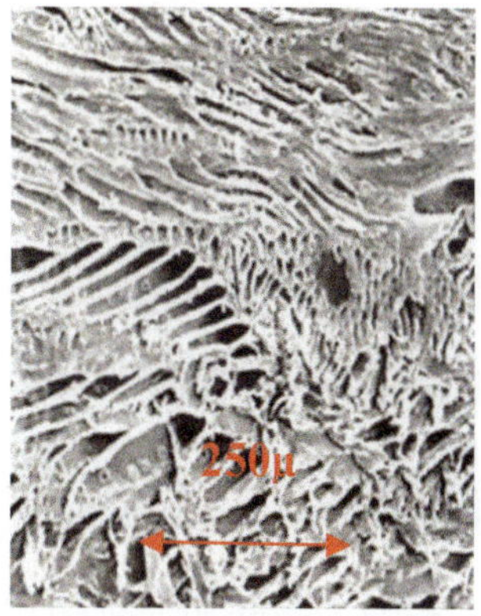

Fig. 3.11 SEM image of abnormally large tracheids in a spruce (*Picea abies*) (Photo: E. Bäucker)

Fig. 3.12 Genetics may be the cause of the abrupt transition from a wide to narrow crown in this spruce (*Picea abies*) (Sweden)

shaped growth of so-called dwarf beech trees (*Fagus sylvatica, var. tortuosa*) (Fig. 3.10). Mutations also account for abnormal cell growth (Fig. 3.11) or can even trigger an abrupt transition from a wide to narrow crown in a spruce tree *(Picea abies)* (Fig. 3.12).

Conclusion: Trees are bound to their genetic specifications. Woodworkers, therefore, must live with their genetic diversity and accept genetic variations in every conceivable form – unless they specifically breed trees to fit their particular needs through *artificial selection* or a *controlled modification of the genetic material.*

3.2 Impact of Physiological Processes Occurring Within the Tree

A tree's vital functions are significantly influenced by the location (climate, soil) on which it grows. The availability of water, nutrients, and light, in particular, determines its internal biochemical processes.

A tree responds to *deficiency symptoms* by altering its growth; for example, if a branch uses up the assimilates it produces itself, instead of exporting them to the stem for radial growth, a moulding will form in the stem section directly below the shade branch (Figs. 3.13 and 3.14).

Air penetration into the stem's interior (branch breakage, internal stem dehydration) can result in oxidative processes

Fig. 3.13 Mouldings in a beech *(Fagus sylvatica)* below a shade branch (Germany)

Fig. 3.14 Mouldings in a bolletrie (*Manilkara bidentata*) (Surinam)

Fig. 3.15 Beech (*Fagus sylvatica*) with facultative red heartwood, visible as cloud heartwood, and pathological rot (black colored)

Fig. 3.16 Oak (*Quercus robur*) with incomplete pith formation

Fig. 3.17 Larch (*Larix*) missing pith leading to "moon rings" (Germany)

Fig. 3.18 Moon ring of unknown cause on kopi (*Goupia glabra*) (Surinam)

Fig. 3.19 Oak (*Quercus* ssp.) burl (bud clusters) (Gran Canaria, Spain)

that lead to facultative heartwood formation. Red heartwood formation is common in beech (Fig. 3.15).

Climatic influences (e.g., early frosts) can prevent cell components from depositing that are needed for pith formation. In such cases, incomplete pith formations (Fig. 3.16), or double sapwood "moon rings" (Fig. 3.17), develop which are distinguishably lighter than the dark heartwood.

The causes of the incomplete heartwood formation in the tropical kopi wood (*Goupia glabra*) shown in Fig. 3.18 are unknown.

Microorganisms can also redirect growth processes in trees to their favor, as easily seen on bumps, burls, and galls (Fig. 3.19).

Conclusion: *A tree can only influence conditions on its growing site over the long term and has no effect on the infectious impact of microorganisms.* Thus, a tree is incapable of preventing any characteristics that they may cause.

Humans can favorably influence a tree's site conditions by improving the soil and through forestry management measures and thereby can gradually alter the physiologically triggered wood characteristics.

3.3 Light/Radiation

The most important influence on tree growth is the photosynthetic effect of direct and diffused radiation at wavelengths between 400 and 700 nm (Promis 2009). A tree is designed to ensure that its assimilation organs, needles or leaves, receive the maximum amount of sunlight. This constant quest for light is called heliotropism. It affects trees in several ways:

1. Leaves and nonwoody shoots react to light variations throughout the day with growth movements or changes in turgor pressure in the leaf stems (Fig. 3.20).
2. Young, woody shoots can adjust to changes in radiation by reorienting themselves through *growth movements*.
3. "Stronger" branches form reaction wood in response to permanent changes in light, appearing as *compression wood* on the underside of softwood branches and *tension wood* on the upper side of hardwood branches (Fig. 3.21).
4. The stem reacts to permanent changes in sunlight exposure by forming reaction wood in its sapwood over the long term. Growth rings or zones widen on the compression-stressed side of the stem in softwoods or on the tensile-stressed side of the stem in hardwoods (Knigge 1958; Mette 1984) (schematic diagrams Figs. 3.22, 3.40, and 3.41).

Fig. 3.20 Linden (*Tilia*), beech (*Fagus*), and ash (*Fraxinus)* leaves optimally positioned for maximum sunlight exposure (aerial canopy view shortly after foliation) (Hainich, Thuringia, Germany)

Fig. 3.21 Branches, crowns, and stems of hedgerow timber adjust themselves with the aid of reaction wood to receive optimal sunlight exposure (Erz Mountains, Saxony, Germany)

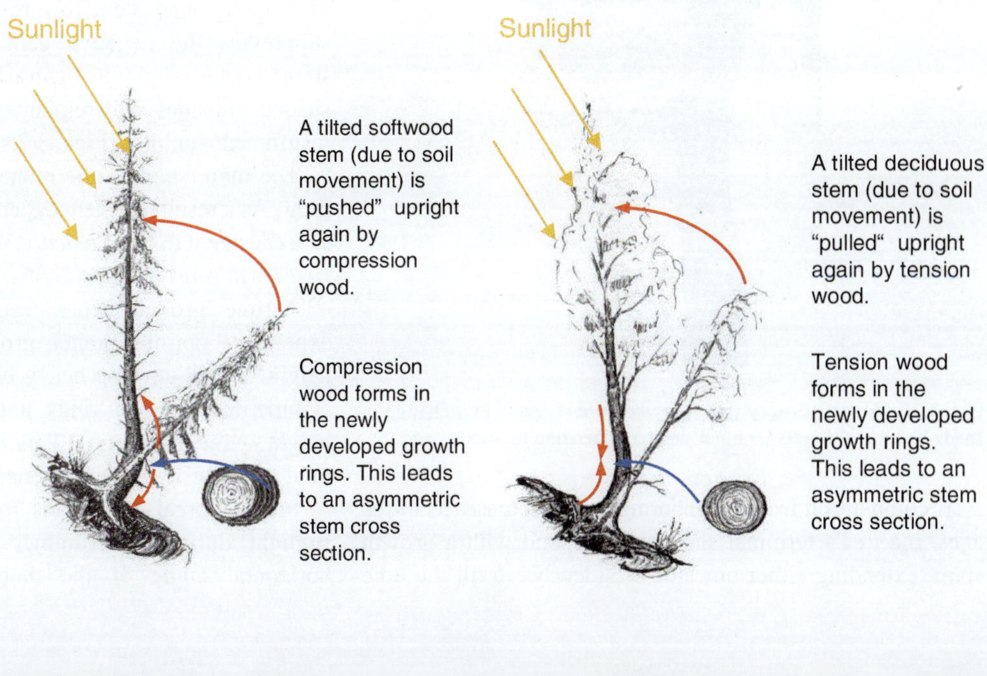

Fig. 3.22 Principle of direction change in the stand: reaction wood formation, appears in softwoods (*right*) as compression wood, in hardwoods (*right outside*) as tension wood with simultaneous modifications to the stem cross section

Fig. 3.23 Heavy competition among coastal redwoods (*Sequoia sempervirens*) leads to conical-crowned, well-formed stems (California, USA)

Fig. 3.25 Red oak (*Quercus rubra*) with stem corrections caused by long-term variations in sunlight exposure

Fig. 3.26 A second-story oak (*Quercus robur*) reacts to a change in shade by reorienting its terminal shoot and bending its stem

If competition from neighboring trees decreases as a tree ages, the tree's terminal shoot will respond with a growth spurt, extending either upwards or sideways to fill the hole

Fig. 3.24 Rode kabbes stem (*Andira* ssp.) in a primary growth forest, solid wood, and branch-free under the crown (KABO, Surinam)

Fig. 3.27 An extreme change in direction of a pine (*Pinus*) branch (loss of apical dominance, negative geotropism, strong heliotropism)

created in the canopy. In this case, phototropism (orientation of branches and stem towards the brightest light source) suppresses the negative geotropism (effort to shift the stems center of gravity) (Strasburger et al. 1978). Heavily shaded branches will remain thin and eventually die from insufficient sunlight (Figs. 3.23 and 3.24).

The more sunlight the branches receive, the thicker they grow. As a result, the stem experiences unequal levels of pressure causing it to build *reaction wood* and leading to *asymmetric growth rings* (Richter 2006a), (Figs. 3.25, 3.26, and 3.27)

A tree growing free from crown competition will experience optimal branch growth and reduced growth in height. It will develop heavy branches and relatively wide, symmetrical growth rings and its stem will taper (low height – diameter ratio). This is how the tree optimizes its vital functions with limited energy expenditure (Fig. 3.28).

In the boreal coniferous forests, the angle of vertical sunlight during the summer is relatively low. Yet the horizontal angle of the daily solar radiation is wide (see Table 3.1). The energy sum from global radiation is below 35 %, in relation to the cloud-free subtropics (Hatzianastassiou et al. 2005). Trees respond by developing low, vertically oriented stress zones and by growing towards the light (Fig. 3.29). Coniferous trees respond to the short vegetation period at low temperatures with an early onset of photosynthesis. Spruce starts to sequester CO_2 at 5 °C (Matyssek et al. 2010).

Fig. 3.28 Western juniper (*Juniperus occidentalis*) with a tapered stem and broad crown due to a lack of competition (Idaho, USA)

Table 3.1 Hours of sunshine, vertical and horizontal radiation based on geographic latitude (cgi Deutschland 2012; Oke 1978)

		Tropics (equator/ Belém) year-round	*Temperate zones* (45° N Lat./Lyon) Vegetation period May–September	*Boreale Zone* (60° N Lat./ Helsinki) Vegetation period June–15 Sept.
Daylight (hours of sunshine)		4.400 h	2.100 h	1.700 h
Vertical radiation direction (vertical angle)	21.3./23.9.	90°	44°	30°
	21.6.	67°	67°	53°
	21.12.	67°	Dormancy	
Horizontal radiation direction (horizontal angle)	21.3./23.9.	180°	180°	180°
	21.6.	133°	240°	290°
	21.12.	227°	Dormancy	

In the deciduous forests of the temperate latitudes, the angle of light during the vegetation period ranges between 44° and 67° (see Table 3.1). The energy sum of global radiation is between 35 and 60 %, relative to the cloud-free subtropics (Hatzianastassiou et al 2005). In the spring, shade leaves in the canopy's upper and lower stories unfold before the sun leaves in the outer edges of the canopy. Since the photoreceptors in shade leaves utilize the dark red portion of the light more efficiently and because shade leaves have a larger specific leaf area (SLA = specific leaf area (m^2/kg)) than leaves (Matyssek et al. 2010), the lower story experiences a temporary lead in assimilation. Given their chloroplast-rich palisade parenchyma, sun leaves probably have a higher CO_2 sequestration rate (Strasburger et al. 1978) than shade leaves, but also age faster. Therefore, a long period of assimilation in the fall lets foliage in the inner canopy and understory compensate, at least partially, for the assimilation deficit experienced during the long summer days. In old growth beech forests, for example, the time difference between sun and shade leaves yellowing may be several weeks (Fig. 3.30). In this way, shade branches and suckers in the temperate latitudes are able to secure their minimum existence.

In old growth tropical forests, the evergreen tree crowns are subjected year-round to sunlight with a steep vertical radiation angle. This leads to strong competition within the canopy (Fig. 3.31). Although the energy sum from global radiation is 60–75 %, in relation to the cloud-free subtropics (Hatzianastassiou et al. 2005), it only barely penetrates the dense, multilayered canopy, preventing shade branches and suckers from developing (Fig. 3.32). This phenomenon does not occur in geometrically arranged tree plantations. In such forests, enough lateral sunlight usually penetrates the canopy to delay the natural pruning processes, thereby increasing the potential need for artificial thinning measures.

Fig. 3.29 The oblique incident of light north of the Arctic Circle leads to heavily branched crowns, especially in spruce (*Picea*) (Rovaniemi, Finland, 68° N Lat)

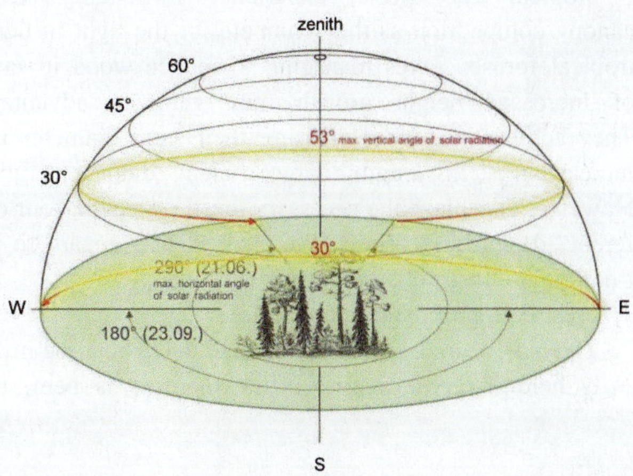

Fig. 3.30 Crown interior, water sprouts, and understory benefit in the temperate latitudes from the early foliage outbreak as well as by the earlier leaf fall in the upper story (Tharandt, Germany, 51° N Lat)

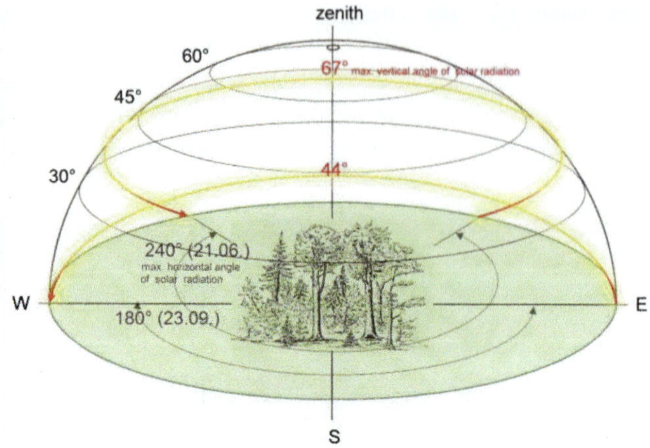

Fig. 3.31 Strong crown competition in the primary growth rain forest results in mushroom-shaped crowns. The only trees with a chance of survival have crown growth that can continue to keep pace with competition (KABO, Surinam)

Depending on geographic proximity to the equator, the steady 12 h of sunshine, the steep vertical angles, and the relatively small horizontal angle of radiation keep shade branches and suckers from living long in dense tropical forests with significant canopy competition.

A model of this relationship is given in the pictures Figs. 3.29, 3.30, and 3.32. Basis for this are the numerical values from Table 3.1.

Tropical tree species experience particularly strong canopy competition as they compete for the light in dense tropical forests. Trees managing to reduce wood in favor of increased height growth may gain an advantage. They achieve this by decreasing their stem diameter and increasing stem length (large height diameter ratio). However, to achieve the necessary tensile strength, four different "design principles" have evolved with regard to the trees' anatomical construction:

1. Prestressing of the trunk (mantle)
   The outside of the tree trunk is held in tension; the inside is held in compression. When the tree is bent, this prestressing usually manages to prevent it from breaking or bulging (principle of prestressed concrete construction). If the stem axis needs to change direction, newly formed tension wood zones along the corresponding section of the trunk make the correction possible (Fig. 3.33).
2. Bandaging the trunk
   As the tree grows in diameter, the fiber directions periodically vary in relation to the stem axis. This alternating spiral effect increases the bending stiffness of the trunk (principle of the crisscrossing bandage) (Fig. 3.34). Direction changes in the stem axis result in the formation of tension wood.
3. Segmentation of the stem cross section
   The stem is reduced to a "construction" of round, positively connected strands of wood (principle of timber frame construction). Tension wood can build on each wood strand in order to correct the direction of the stem axis (Fig. 3.35).
4. Stabilization of the stem base
   Wide-spreading buttress roots form at the base of the stem (principle of foundation enlargement) improving the stability of the trunk (Fig. 3.36).

These four "design principles" are often combined. They can also be found to a lesser degree in trees from temperate climates.

Conclusion with Regard to the Impact of Light/Radiation: Trees can react to changes in light and radiation by forming reaction wood. This explains many of the different wood characteristics, especially those found in the tree trunks (curvature, unroundness). Depending on the amount of crown competition and the geographic latitude, sunlight and vertical and horizontal radiation towards of the sun influence branch growth and the formation of water sprouts on tree stems.

Over the course of their evolutionary history, trees have adjusted to heavy crown and light competition by developing *material-saving stem constructions that benefit crown and root growth.*

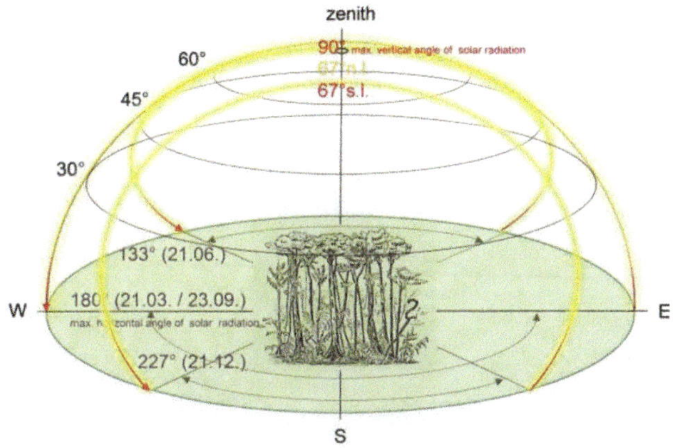

Fig. 3.32 The steep angle of light incidence and strong crown competition in the tropical primary growth forests prevent water sprouts and low-setting branches (Witagron, Surinam, 4° N Lat)

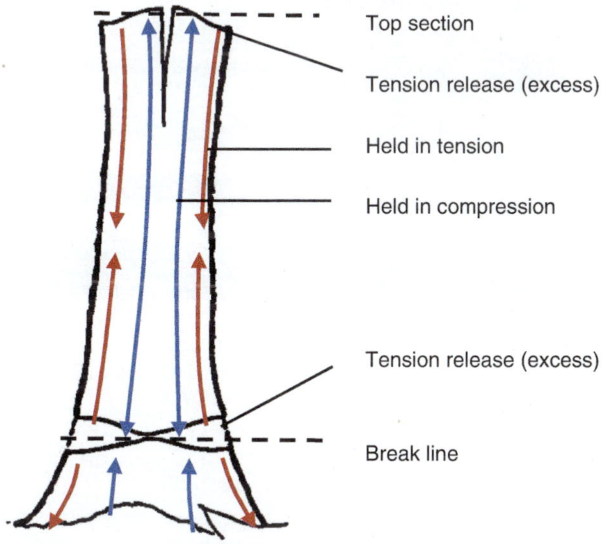

Fig. 3.33 Growth stresses in hardwood: tensile stress in the sapwood and compression stress in the stem core. When a tree is felled and the internal tension is equalized. The resulting decompression inside the stem can lead to stress cracks (shakes)

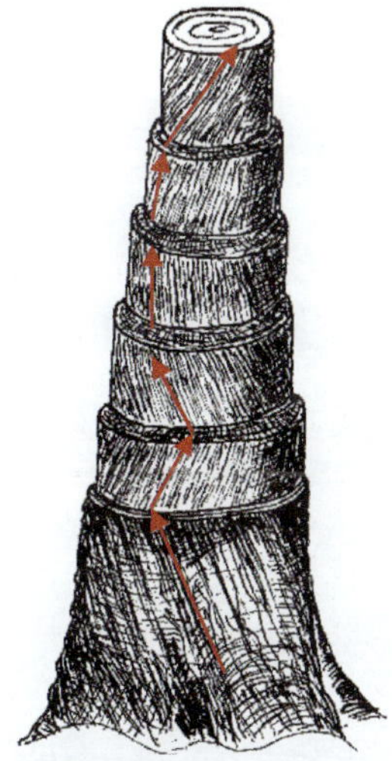

Fig. 3.34 Periodically alternating spiral grain. The overlapping fiber structure makes the stem difficult to split after it is felled

Fig. 3.35 The stem is formed from interlinked, bar-shaped strands

3.4 Mechanical Stress

When external forces impact the side of a tree or tilt the stem from its vertical position (wind exposure, snow buildup, soil movement, lopsided crowning), the tree responds by forming *reaction wood.* Reaction wood counteracts the external pressures and enables the tree to orientate its stem back to the light (phototropism) source or against gravity (negative geotropism) and to better distribute the mechanical stress on the stem or branch (Knigge 1958; Mette 1984) (Figs. 3.37 and 3.38). Mattheck (1997 p. 14) describes this effect as the "axiom of constant tension," although this actually only applies to isotropic bodies and not to a living tree as an anisotropic body (Sinn 2009).

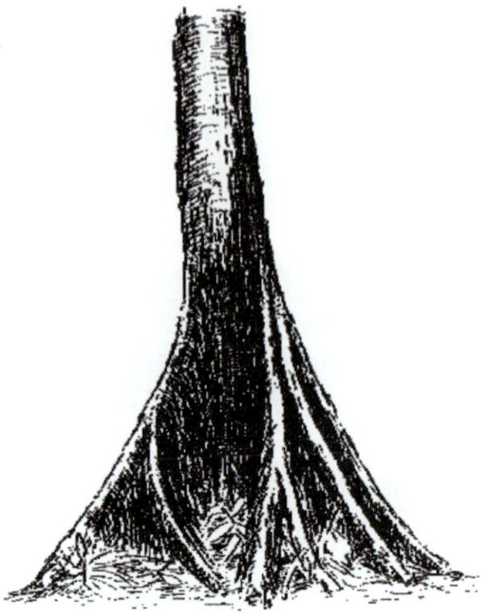

Fig. 3.36 Buttress roots improve the stability of this solid wood stem

Fig. 3.37 Wind stressed Douglas firs (*Pseudotsuga*) straighten themsleves by forming compression wood

Fig. 3.38 Leaning djadidja (*Sclerobium melinonii*) stem counteracting gravity with extreme tension wood formation djadidja (*Sclerobium melinonii*), Surinam

Fig. 3.39 Beech (*Fagus*) tilted by steady winds blowing across the North Sea Dike. The lateral stress superseded the geotropic forces and partially overrode the effects of heliotropism, Lower Saxony, Germany

Extreme mechanical stress applied horizontally is capable of superseding the forces of negative geotropism and partly outweighing the influence of heliotropism (Fig. 3.39).

According to Rosenthal (2009) and Rosenthal and Bäucker (2012), the alignment of microfibrils in the cell wall is key to the formation of both compression and tension wood. In accordance with the lignin swelling theory, lignin molecules fill the available spaces between the already existing microfibril. As a result, compression stress occurs at a right angle to the direction of the microfibrils. If the

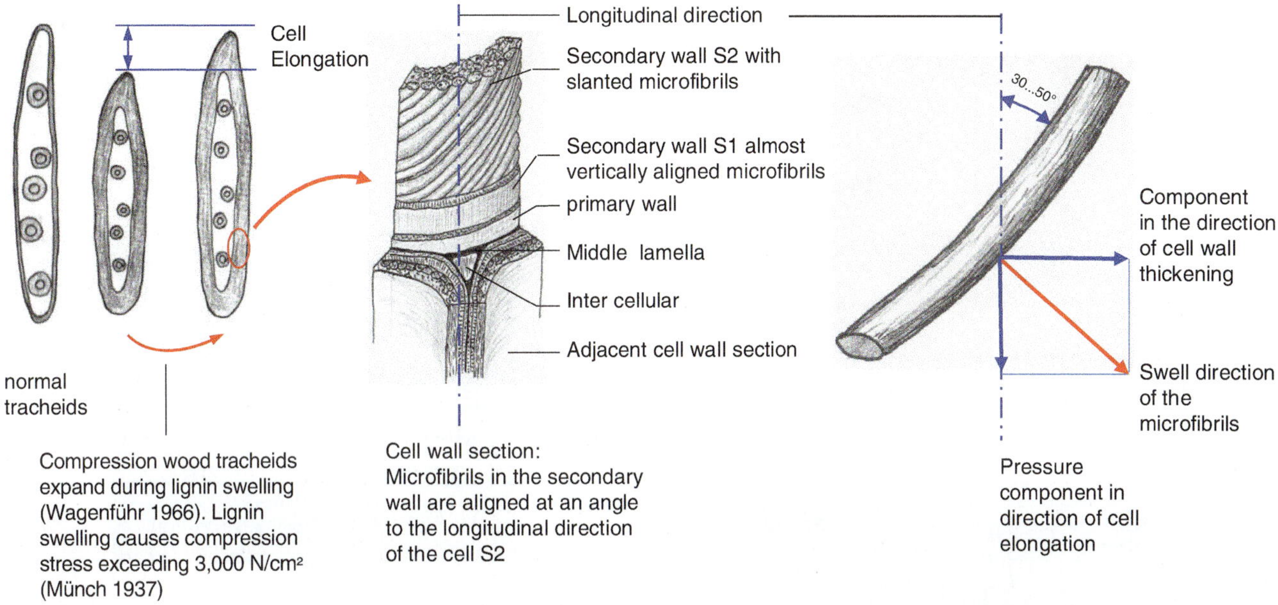

Fig. 3.40 Principle of length variation in the compression wood of softwoods

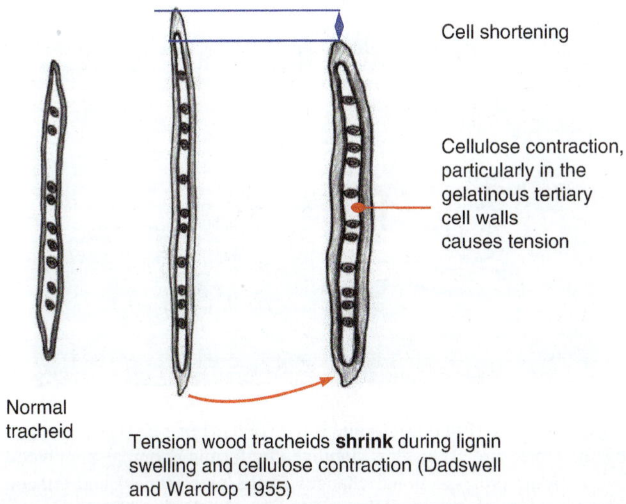

Fig. 3.41 Principle of length variation in tension wood of hardwoods

Fig. 3.42 Extremely eccentric growth at the stem base of a pine (*Pinus*) after decades of unilateral wind stress (Germany)

microfibril angle – as typical for compressed wood – lies between 30° and 50°, based on the cell's longitudinal direction, then the compression stress caused by the lignin swelling will lead to an extension of the cell wall; *compression wood* forms (Fig. 3.40).

If the microfibrils are not, or only slightly, angled in the longitudinal direction of the cell, then the compressive stress caused by the swelling of the lignin will lead the cell walls to thicken. In addition, a highly soluble gelatinous substance deposited in the wood fibers shortens the cell walls during swelling. This leads the cellulose to contract (Matyssek et al. 2010) and *tension wood* forms (Wagenführ 1966) (Fig. 3.41).

Softwoods form *compression wood* on the lower, compressed side of leaning branches, root collars, and stems. Tracheids in compression wood are relatively short and rounded with significantly thicker cell walls. A high level of lignin gives the growth rings on the compressed side a reddish brown color, making it difficult to identify the transition from early to latewood tracheids. The tendency to form compression wood differs between softwood species (Dadswell and Wardrop 1955), (Figs. 3.42 and 3.43).

Hardwoods form *tension wood* on the upper side of the lean (Fig. 3.44). Tension wood is difficult to identify. In manufacturing, tension wood often produces wood with a "woolly" surface (Fig. 3.45). Tension wood can be identified

Fig. 3.43 82 years of unilateral wind stress led to extreme compression wood formation in this juniper (*Juniperus communis*) (Finnmark, Finland)

Fig. 3.44 Tension wood pulls a beech (*Fagus*) upright after it is slanted by ground movement in sedimentary soil (Northern Limestone Alps, Germany (Nördliche Kalkalpen))

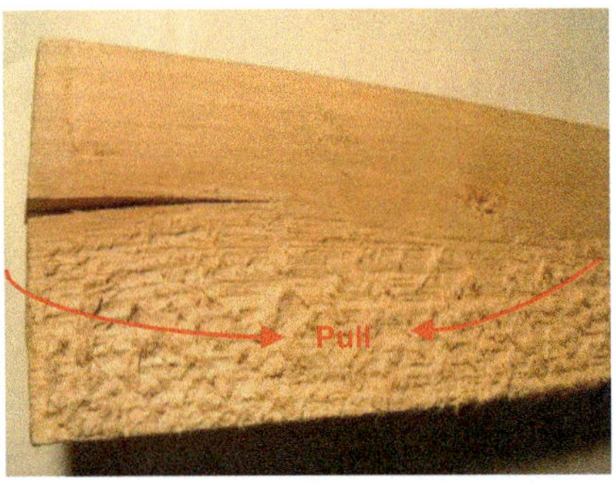

Fig. 3.45 Tension wood in a willow board (*Populus Salix alba*) with woolly fibers and stress crack

Fig. 3.46 Extreme reaction wood formation in a hoogland bebe (*Pterocarpus rohrii)* in tropical primary forest (Surinam)

Fig. 3.47 Pine (*Pinus*) (*left*) and birch (*Betula*) (*right*) crowns broken by snow pressure. The stems stabilized by forming compression wood and extreme cross-sectional changes in the pine (*Pinus*) and tension wood in the birch (*Betula*) (Finnmark, Finland) (Richter 2003b)

chemically based on its high cellulose content with Astra blue. Among hardwoods from the temperate zones, beech (*Fagus*), poplars (*Populus*), oaks (*Quercus*), and elms (*Ulmus*) are particularly prone to forming tension wood (Mette 1984).

Tropical hardwoods also form tension wood on sections of the stem that rely on short-term reinforcements in cases of gravitational crown shifts, particularly in primary forests (Fig. 3.46).

Reaction wood in phylogenetically older softwoods places the wood fibers under compressive stress and places phylogenetically younger hardwoods under tension. Because the compressive strength of wood fibers is only about half of its tensile strength, about twice as much compression wood forms compared to tension wood under the same amount of stress. The difference in levels of reaction wood formation between a less stressed hardwood stem and a softwood stem can be seen in the stem cross sections (Fig. 3.47).

Hardwoods are capable of reorienting themselves with less material expenditure than softwoods. For example, this ability is extremely pronounced in the tropical tree species witte parelhout (*Aspidosperma marcgrafianum*): The typically round- to oval-shaped stems become fluted. Given the rapid changes in tensile stress, increment zones are only able to form in certain sections of the stem (Fig. 3.48).

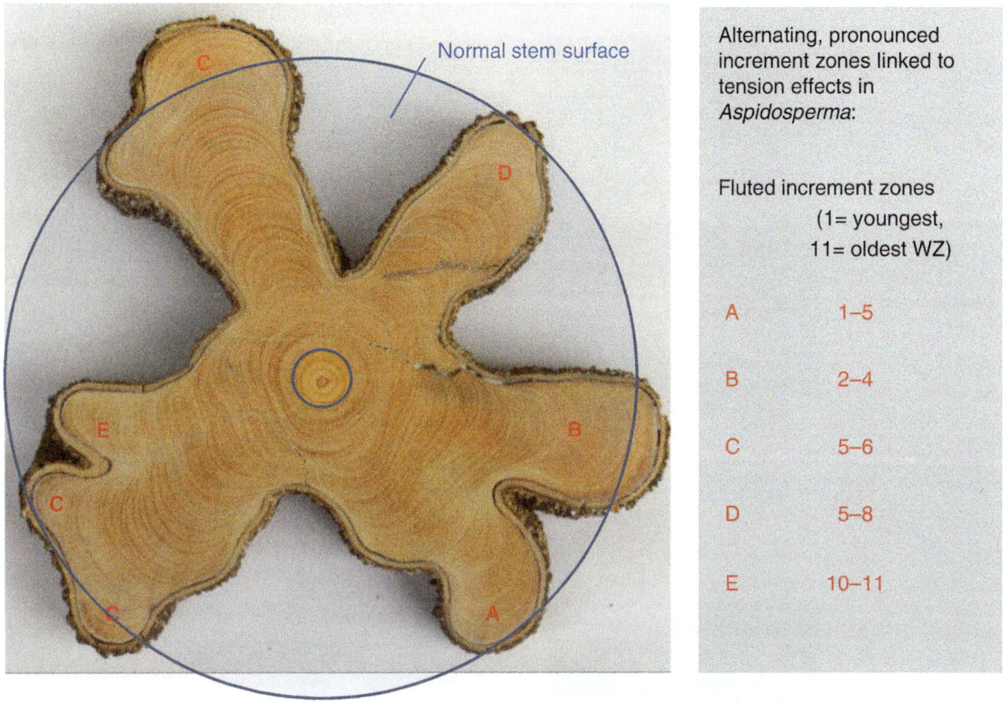

Fig. 3.48 Tension wood caused by sudden repositioning of the stem and crown leads to fluting in the tropical tree species witte parelhout (*Aspidosperma marcgrafianum*) (Surinam)

Conclusion to the Influence of Mechanical Stress: Trees are capable of actively reacting to mechanical stress by forming reaction wood.

This explains many wood characteristics, especially those which can be identified from the trees' outward appearance (curvature, unroundness, cracking).

3.5 Injury

Trees react to stem injuries in the short term by producing *wound closure material*. These exudates appear as water-insoluble resin in softwoods as well as in tropical species (Fig. 3.49). Tropical species are particularly adept in protecting wound surfaces with water-soluble or water-retaining gums, resin-based kinos, or polyterpene-based lattices (Lange 1998a, b, c) (Fig. 3.50).

At the same time, trees respond internally to a stem injury or a broken branch by *walling off* the healthy tissue (compartmentalization) as follows (Dujesiefken and Liese 2006):

Phase 1: Air penetrates into the injured tissue and dries it out. Accessory components (chemical substances) deposit in the wood boundary layer of the dried tissue. Wound periderm forms to protect the water-conducting cells. In hardwoods, peripheral vessels with tyloses form. In softwoods, boarder pits close off any tracheids that are still intact. Softwoods with resin ducts respond by accumulating resin. A callus forms around the wound in an effort to seal off the area.

Phase 2: Wound periderm and wood boundary layers prevent harmful pathogens from further penetration.

Phase 3: Pathogens spread through the wood. If a wood boundary layer is penetrated, more boundary layers can form in which accessory components are again deposited.

Phase 4: The wound is fully covered from the vertical wound boarder with *wound wood*; the pathogens are encapsulated (Fig. 3.51).

Wound wood differs anatomically from the normal wood in that it generally lacks wood fibers, libriform fibers, and tracheids. Instead, it forms thickened parenchyma cells. As a result, the wound wood is not as hard (Sinn 2009).

The healthy tissue initially walls itself off from air and later from microorganisms to different degrees in the longitudinal, tangential, and radial direction, as well as along the boundary of the newly formed wound wood (Fig. 3.52).

According to the CODIT model (*Compartmentalization of Damage in Trees*) by Shigo (1990), compartmentalization progresses as follows:

Zone 1: Minimal vertical compartmentalization occurs along the vascular system or tracheids (a few protective transversal cell walls).

Zone 2: Moderate tangential compartmentalization internally, along the growth rings (many pit transitions to adjacent cells).

Zone 3: Good radial compartmentalization (relatively few ray cells).

Zone 4: Very good tangential compartmentalization externally. This barrier protects the new tissue that developed after the wound from fungi.

The effectiveness of the compartmentalization process varies from tree species and depends on the season in which the injury occurs. Compartmentalization is generally more successful in the spring and late summer than in a tree's

Fig. 3.49 The resin groove in the pine stem (*Pinus*) has been callused over for 15 years. The wound is walled off, isolating it from infection (Brandenburg, Germany)

Fig. 3.50 The fresh wound on this bolletrie *(Manilkara bidentata)* is immediately sealed with latex milk (Surinam)

Fig. 3.51 An injury on an oak (*Quercus*) is walled off sideways by wound wood isolating the infectious area

dormant periods (Dujesiefken and Liese 2006). Optimally, the wood in the wounded area is simply discolored, while the fiber tissue remains mostly intact (Figs. 3.53 and 3.54).

Under tropical climatic conditions, stem injuries and branch breaks caused by consistently severe infection stress often have serious consequences (Fig. 3.55). In the worst-case scenario, the wound can lead to such extensive decay that the tree is unable to compartmentalize (Dujesiefken and Liese 1990, 2011; Dujesiefken et al. 1991).

A tree's radial growth and the advancing decay are in direct competition. Trees with such extensive injury no longer add value to a stand (Fig. 3.56).

A special type of wood wound with a similar anatomical structure is the so-called "*Wulstholz*" (bead wood) forms after compression failure, when the stem fibers are compressed beyond their limits (Figs. 3.57, 3.58, 3.59, and 3.60). Like compression wood, the cell walls of Wulstholz have a very wide microfibril angle, associated with reduced stiffness (Trendelenburg 1941; Rosenthal 2009). Compression failure and fiber fractures appear in trees from the temperate zones as well as in the tropics. Wulstholz can also form as a frost crack (Fig. 3.61) or as callused lighting groove (Wagenführ 1966). The Wulstholz seals off the wound from harmful pathogens, but does not increase the tree's stability.

Conclusion to Stem Injury: Trees are able to respond actively to stem injuries by producing resinous exudates and forming wound wood or "Wulstholz." *Internally, the injuries are walled off through compartmentalization.* This explains many wood characteristics, especially those visible on the stem (unroundness, bulge), discoloration, and decay.

3.6 Summary of the General Factors that Lead to the Formation of Wood Characteristics

Trees cannot actively respond to the two "internal" factors that create wood characteristics, *genetic predisposition* and *genetic modification*, nor to the *physiological conditions* to which they are subjected. The wood characteristics they trigger are unavoidable. Trees do, however, have two effective means of actively responding to the three "external" factors light, mechanical stress, and injuries:

1. They react to light and stress factors by forming *reaction wood.*
2. They wall off injuries externally with *exudates* and *wound wood* or *Wulstholz* and internally through *compartmentalization.*

The following first introduces the principal "triggers" that cause wood characteristic to form then gives a comprehensive overview of the main characteristics, and finally, Part III, discusses how the main triggers affect the development of the individual wood characteristics.

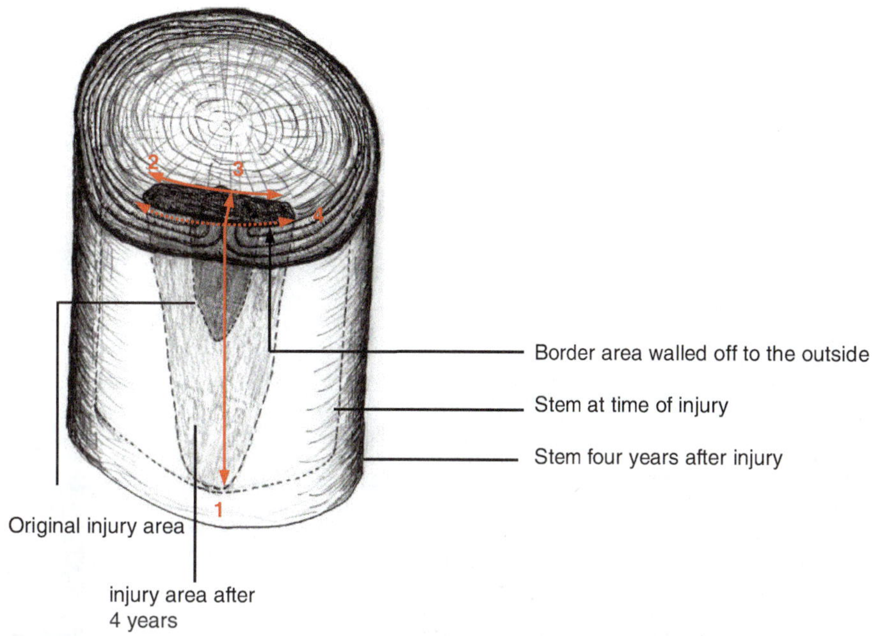

Fig. 3.52 Wound wood walls off an injury to different degrees at the various steps of compartmentalization (Shigo 1990). *Levels of compartmentalization in different directions*: *1* longitudinal weakest wall, *2* tangential, along the growth rings' second weakest wall, *3* radial, along the rays, strong wall, *4* wound wood between wound and callus, strongest wall

Fig. 3.53 A logging wound in this beech (*Fagus*) is walled off by wound tissue and compartmentalized. Red heartwood forms behind the wound (oxidative discoloring and tyloses formation through drying)

Fig. 3.55 12 years after a logging injury (only bark damage!), all wound wood sections have broken open in this tropical rain forest stem (Surinam)

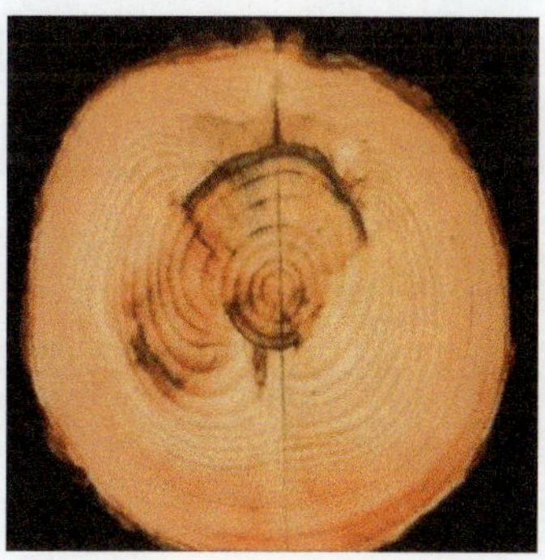

Fig. 3.54 The bark damage in this spruce (*Picea*) is completely walled over

Fig. 3.56 Red rot (*Heterobasidion annosum*) only affects the dead heartwood of this spruce (*Picea*) (*left*). Because the red rot was not successfully walled off, the tree can only survive through strong radial growth (bottle-shaped growth, *right*)

Fig. 3.57 Wulstholz 5 years after a storm in a 70-year-old spruce stand (*Picea*) (Germany)

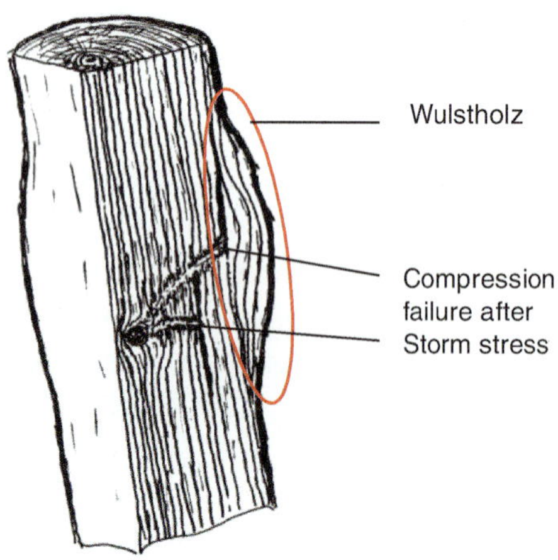

Fig. 3.58 Wulstholz walls off compression failure as a reaction to abiotic external stress

Fig. 3.59 Compression failure in the base of a zwart riemhout (*Micropholis guyanensis, var. commixta*) (Surinam)

Fig. 3.60 The fiber deformation in this spruce (*Picea*) was covered over by elastic Wulstholz

Fig. 3.61 Ash (*Fraxinus*) with frost crack sealed with Wulstholz

Part II

Wood Characteristics Overview

4 Overview of the Main Wood Characteristics

Wood characteristics are categorized into three characteristic groups along with an overview of the crack forms and causes:

The first grouping covers *wood characteristics that form as a tree grows naturally*. These characteristics are either genetically fixed or physiologically determined and develop naturally as a tree grows. For example, as every tree forms branches to transport assimilates, it also responds to light stimuli, site and climate influences, modified nutrient supply, external forces and stress. The tree stem adapts by deviating from its normal contour. Branches respond by either growing stronger or dying off. Changes may also occur in the direction of the fiber, tree ring structure or increment zone formation, and color of the wood. Humans can influence some of these characteristics through forest management practices.

Important wood characteristics, particularly those visible on standing timber or felled logs, will be highlighted in this chapter with two stars (**). Wood characteristics not specifically covered in this chapter, but still mentioned in the descriptions, have one star (*).

C. Richter, *Wood Characteristics: Description, Causes, Prevention, Impact on Use and Technological Adaptation*,
DOI 10.1007/978-3-319-07422-1_4,

Wood characteristics inherent to a tree's natural growth

- **Stem contour modifications**
  - Taper**
  - Crookedness**
  - Forking**
    - Fused stem
    - True forking
  - Out-of-roundness**
    - Ovality
    - Eccentric growth
    - Mouldings
    - Fluting/flanges
- **Limbiness**
  - Branches**
    - Living branches
    - Dead branches
    - Suckers/epicormic shoots
  - Branch scars**
    - "Roses" (coarsely barked tree species)
    - "Chinese´s mustaches" (smooth barked tree species)
    - Branch bump
- **Anatomical structure**
  - Irregular growth ring formation**
  - Fiber orientation**
    - Spiral growth
    - Wavy growth
    - Fiddleback
    - Hazel growth
  - Reaction wood**
    - Compression wood in softwoods
    - Tension wood in hardwoods
  - Wounds/"Wulstholz"**
  - Ingrowths
    - Resin pockets*
    - Bark pockets*
    - Mineral pockets
  - Growth stresses/stress cracks*
- **Color changes**
  - True heartwood*
  - Facultative heartwood*
    - Red/grey/olive/brown heart*
    - Irregularly formed heartwood*

The second group of wood characteristics comprises *biotically induced characteristics*. These include microorganisms and animals that use tree parts as a food source or for nesting. Human influences include injuries due to forestry or logging operations, as well as damage caused by warfare, carelessness, malicious intent, or special interest groups. In the tropics, parasitic or saprophytic plants and plants which use the tree for climbing support have the greatest impact.

With appropriate intervention, humans can minimize and sometimes even prevent the harmful influences of microbes, insects, and animals on the wood. Education and a sound understanding of effective forest management practices can limit the threats posed by humans.

In tropical primary and secondary forests, human "corrective" influence on the vulnerable ecosystems is always problematic.

Biotically induced wood characteristics

- **Effect of viruses, bacteria, and fungi**
  - Soft rot*
    - Blue stain
    - Red striped
  - White rot*
    - Red rot in spruce
    - Tinder fungus
  - Brown rot*
    - Pine tree rot
    - Honeycombing
    - Honey fungus
  - Rust fungi*
  - Necroses**
  - Cancer**
    - Burls/bark burls
    - Galls
    - Witches' brooms
- **Effect of animals/humans**
  - Molluscs
  - Insect damage*
    - Damage by bark nesters
    - Damage by wood nesters
  - Vertebrates**
    - Birds
    - Browsing
    - Rubbing
    - Peeling
    - Gnawing
  - Forestry operations**
    - Felling
    - Hauling
    - Resin extraction
  - Stem injuries**
    - Warfare
    - Carelessness/malicious intent
    - Sports clubs/hunting
    - Special interest groups/nature Conservation
- **Effect of plants**
  - (Semi -) parasites**
  - Climbers/twines/root climbers**

Abiotically induced wood characteristics

- **Temperature** -humidity -effect
  - Bark scorch**
  - Dry crack*
  - Suction tension crack*
  - Lightning**
    - Lightning moulding
    - Lightning hole
  - Forest fire*
  - Frost crack/frost scar**
  - Frost heart*
  - Moon ring*
  - Hail
  - Water damage
- **Wind and snow effects**
  - Compression failure/fiber fractures**
  - Shear stress crack**
  - Branch demolition
  - Stem break*
- **Rock fall***

Crack forms with varied causes**

- Heart shake, cross crack, star shake
- Traversing shake
- Ring shake
- "Spider" cracks
- Radial shake
- Tangential crack/"schilfer"shake
- Fiber fracture

The third group of wood characteristics includes abiotically induced wood characteristics of inanimate nature. Temperature, precipitation, lightning, wind, and snow cannot be easily influenced. Humans can, however, minimize some of their potential damage through preemptive forest management practices that include selecting tree species suitable for a specific site and planting stands and individual trees with appropriate spacing.

In tropical rainforests, abiotic injuries are usually the result of heavy rainfall associated with strong storms.

The descriptions of the cracks can sometimes be relatively imprecise because there is no exact distinction between the actual crack *forms* and the underlying *crack causes*. On the one hand, some crack forms can have several different causes. For example, a cross crack could be caused by dehydration or by growth stresses. On the other hand, the same causes can lead to several different crack forms. Growth stresses, for example, may lead to cracks in the cross-section area, cross cracks, or star shakes. These connections will be clearly illustrated in the summary on crack forms.

Part III

Description of the Wood Characteristics

5 Wood Characteristics Inherent in a Tree's Natural Growth

5.1 Stem Contour Modifications

5.1.1 Taper (Richter 2002a)

5.1.1.1 Description

Taper refers to a progressive reduction in a stem's diameter from base to tip. The height-diameter relationship (*H*/*D* ratio) significantly influences wood volume recovery. Tapered stems produce markedly more slab and edge wood when processed. Either the sawed lumber has the desired diameter, but insufficient length, or it has the right length, but too small a diameter. This dilemma has challenged wood workers throughout the ages. Early timber traders established grading methods based on top diameter and length as first documented in the *Gengenbach Sawmill Ordinance* of 1430 and the *Württemberg Raft Ordinance* of 1588 (Willing 1989).

Around 1900, a model eventually developed in Southern Germany that classified timber based on minimum length and minimum top diameter called the Heilbronn Grading Rule (Table 5.1). In the meantime, a system based on volume, called mid-diameter grading, became standard in Northern Germany. Today, most timber in Germany is classified using this mid-diameter grading rule.

Measurement: The "Framework Agreement for Timber Trade in Germany" (RVR 2014) determines stem taper by measuring stem diameter *D* at 1 m (for butt logs) and *d* at least 5 m from both ends of a log. The reference length *L* lies between the two measurement points. This method excludes the stem butt from the calculation, but not the heavily tapered crown (Fig. 5.1):

$$\text{Stemtaper}\,[\text{cm / m}] = (\text{D} - \text{d}) : \text{L}$$

Based on the "Timber Grading Rules in Germany" (HKS 2002a), diameter is measured at points 1.0 m inwards from both ends. The European Committee for Standardization (CEN) (DIN 1999a), however, sets the measurement points at "at least 5 cm" inwards.

Timber is classified as heavily tapered if the diameter of the stem decreases more than 1 cm for each meter.

See Plate 5.1 for stem taper photos.

Table 5.1 Spruce (*Picea abies*), fir (*Abies alba*), and Douglas fir (*Pseudotsuga menziesii*) measured using Heilbronn Grading Rules (Helm 1982)

Class	Min. length (m)	Min. top diameter (cm)
H 1	8	10
H 2	10	12
H 3	14	14
H 4	16	17
H 5	18	22
H 6	18	30

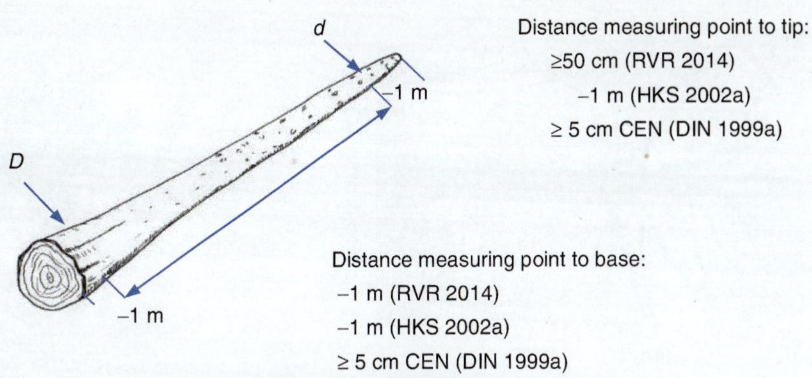

Fig. 5.1 Measurement of a tapered spruce (*Picea*) butt log according to various grading standards

C. Richter, *Wood Characteristics: Description, Causes, Prevention, Impact on Use and Technological Adaptation*, DOI 10.1007/978-3-319-07422-1_5, © Springer International Publishing Switzerland 2015

Tapered oak crown (*Quercus robur*)

Dawn red wood (Metasequoia glypto stroboides) predisposed to taper

Two old dead ponderosa pines (Pinus ponderosa) on the same site but with different stem forms: *left* tapered, *right* well formed (Utah, USA)

Tapered spruce at timberline (Lapland, Finland)

Extreme taper on a Siberian larch (*Larix sibirica*) (Hubsugul Nur, Mongolia) (Photo: M. Bürger)

Free standing juniper (*Juniperus occidentalis*) with heavy taper (Idaho, USA)

Well-formed stems in tropical primary and secondary forest due to heavy crown competition. Buttress roots often form to stabilize the narrow stems. Right basal section of a bosmahony (*Martiusia parviflora*) (Surinam)

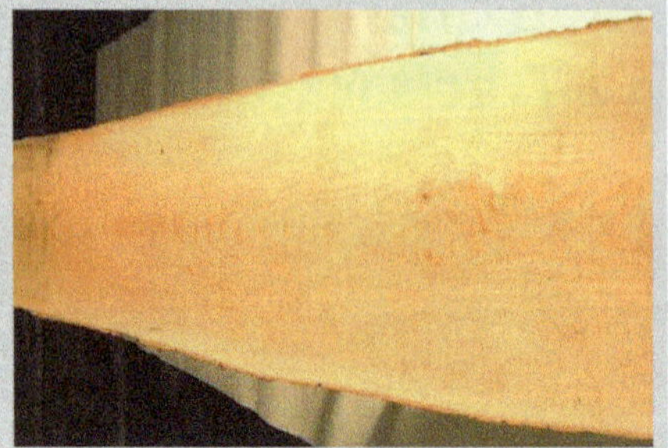

Heavily figured birch (*Betula*) board from a tapered stem section

Plate 5.1 Stem taper

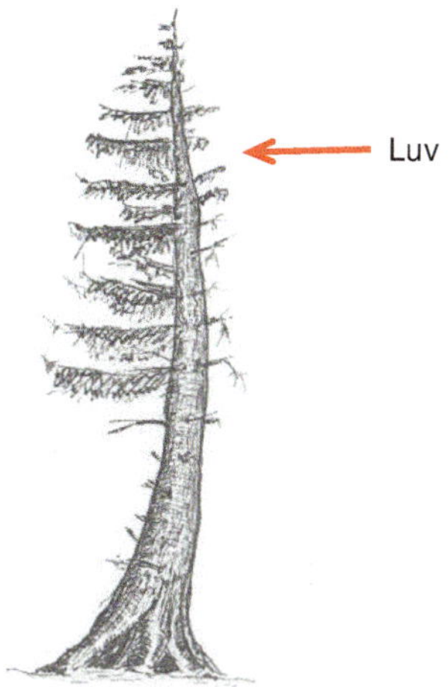

Fig. 5.2 Spruce (*Picea abies*) on exposed cliff site

5.1.1.2 Causes

Certain tree species, for example, hornbeam (*Carpinus betulus*), yew (*Taxus baccata*), swamp cypress (*Taxodium*), and arborvitae (*Thuja occidentalis*), are predisposed to stem taper.

As the density of trees in a forest stand increases, crown competition rises and the individual trees strive to grow taller and break through the canopy. This growth in height takes place at the expense of growth in diameter and stand stability. Stands without crown competition promote trees with broad crowns. The trees improve stability by increasing cambial growth.

Low density stands promote broad crown growth. This results in low *H*/*D* ratios. The same applies to solitary stems with low-set crowning. The trees grow faster in width with reduced height growth.

Most species that experience slow growth in height form tapered crowns.

On unfavorable sites, highly influenced by wind or unstable soil conditions (cliffs, mountain ridges, stand edges, or bogs), trees improve their stability by increasing cambial increment growth in their lower stem (low *H*/*D* ratio) (Figs. 5.2 and 5.3).

As a general rule, the younger and more stable the tree, the greater the crown competition, the more narrow the stem ($H/D>80$). The older and less stable the tree, the lower the crown competition, the thicker the stem ($H/D<50$) (Rust et al. 2011).

Physically, stem taper can be understood as an increase in the stem base diameter of a tree resulting in a decrease in its bending tension to the third power over trees of the same height (Mattheck 1997):

$$\sigma = \frac{4M}{\pi \times R^3}$$

Legend: σ = bending stress, M = bending moment ($F \times L$), F = force on the crown, L = load (stem length), R = diameter at stem base

Fig. 5.3 Swamp cypress (*Taxodium*) on boggy soil. External forces and site conditions affect the *H*/*D* ratio

Fig. 5.4 Spacing affects the *H*/*D* ratio

That means, for example, if a stem exposed to steady wind pressure doubles its diameter, it reduces its bending stress by 1/8. Consequently, the tree increases its chances of survival.

In tropical primary and secondary forests, trees normally have only slight tapering ($H/D>80$). As growth in height accelerates to overcome crown competition, the stem diameters and crown widths progressively decline. Over time, various "design principles" have evolved to ensure stability of the trees (see Sect. 3.3).

5.1.1.3 Prevention

Forest management practices such as suitable species selection and appropriate spacing can influence a tree's *H*/*D* ratio (Fig. 5.4).

5.1.1.4 Impact on Use

The wood volume yield from sawn logs is reduced because taper leads to shorter board lengths or widths. The amount of slab and edge wood (offcuts) increases (Fig. 5.5).

Missing fiber in the sawn timber reduces the strengthening properties. A variance in fiber of 5° to the surface of a board reduces the bending strength by 20 %. At a 10° fiber angle, bending strength is reduced to a critical 40 % (Fig. 5.6) (Pope et al. 2005).

The surface quality declines when the wood is planed against the missing fiber (Fig. 5.7).

The effect of stem taper on grade quality among conifers is quantified based on the "Framework Agreement for Timber Trade in Germany" (RVR 2014) in Table 5.2.

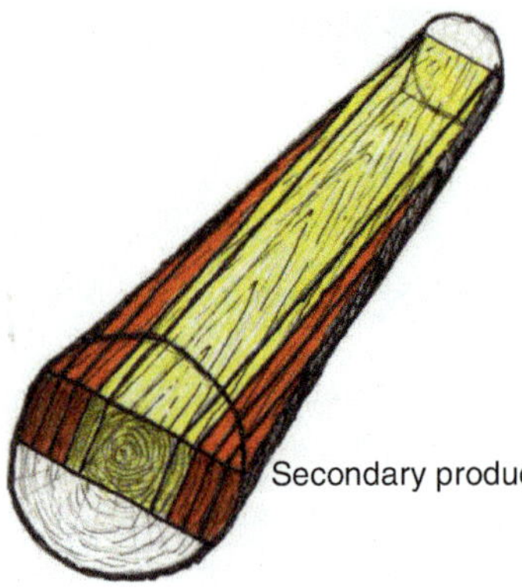

Fig. 5.5 Flitch cut from a highly tapered log: loss in volume recovery from tapered logs. Wide and thin ends with missing fiber

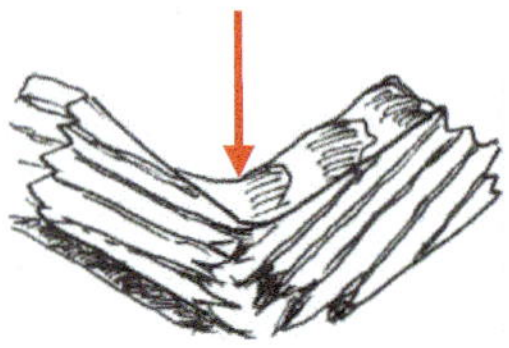

Fig. 5.6 Reduced bending strength due to missing fibers

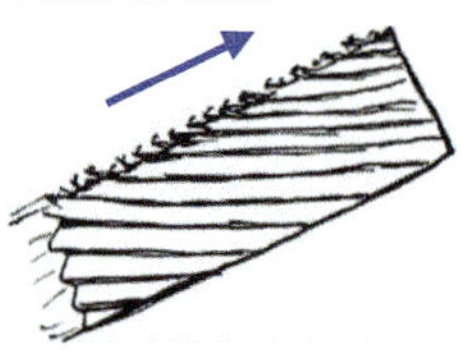

Fig. 5.7 Rough surface caused by planing against the grain

5.1.1.5 Technological Adaptation

Modern band saw technology has optimized the processing of tapered logs. The wood is cut parallel to the stem surface in the direction of the fibers, conical profile (Fig. 5.8).

Round-cut boards can be trimmed parallel to the wane. These trimmings eliminate waste and are sufficient for veneering wood (roof boards) (Fig. 5.9).

Forest measurements collected manually can be time consuming and inaccurate. Therefore, most sawmills today sort timber using (semi)automated optoelectronic scanners. These systems are able to scan the stem contours – dependant on the desired log length – in many evenly distributed sections. The scanners are calibrated and ensure measurements in line with accepted grading rules.

5.1.2 Crookedness (Richter and Mahler 2003)

5.1.2.1 Description

Crookedness refers to stem deviation from a straight line along the longitudinal axis. A straight stem is called "double lined," a stem curving to one side "single lined or sweep," and a stem curving at different stem heights "unlined or crook" (Fig. 5.10).

The term "lined" comes from the plumb line. When a plumb bob is suspended down the length of a standing tree, the plumb line falls either in line with the stem (lined), or out of line (unlined), or across the crooked stem section.

Throughout history, woodworkers have consistently found ingenious ways to use crooked stem parts. By selecting wood with naturally suitable shapes, they created everyday objects that were inherently stronger and required less material to produce. Stone Age digging sticks and ax handles provide the first evidence of such practical tools. The use of naturally curved wood continued well into the Middle Ages as seen in the period architecture and flourish of woodworking professions, such as wheelwrights and coopers. Demand for naturally shaped timber experienced a heyday in the late Middle Ages when large-scale shipbuilding, particularly in France, England, Holland, Portugal, Spain, and Venice, devoured vast

Table 5.2 Effect of stem taper on grade quality among conifers (RVR 2014)

Species	Average diameter (cm)	RVR quality class			
		A	B (cm/m)	C (cm/m)	D
Spruce/fir (*Picea abies/Abies alba*)	Ø<20	–	≤1.25	≤2	Unlimited
	Ø≥20<35	Unlimited	≤1.5	≤2.5	
	Ø≥35		≤2	≤4	
Pine (*Pinus sylvestris*)	Ø<35	Unlimited	≤1.5	≤2.5	Unlimited
	Ø≥35		≤2	≤4	
Douglas fir (*Pseudotsuga menziesii*)	Ø<35	Unlimited	≤1.5	≤2.5	Unlimited
	Ø≥35		≤2	≤4	
Larch (*Larix decidua*)	Ø<35	Unlimited	≤1.5	≤2.5	Unlimited
	Ø≥35		≤2	≤4	

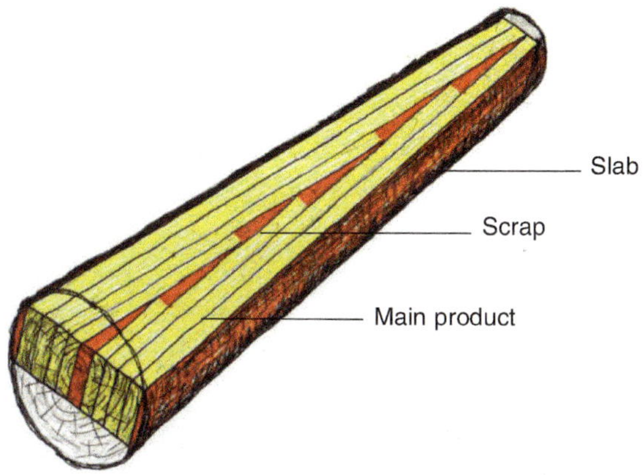

Fig. 5.8 Conical cut diagram using a band saw. Wide end of board parallel to fiber, narrow end with missing fiber

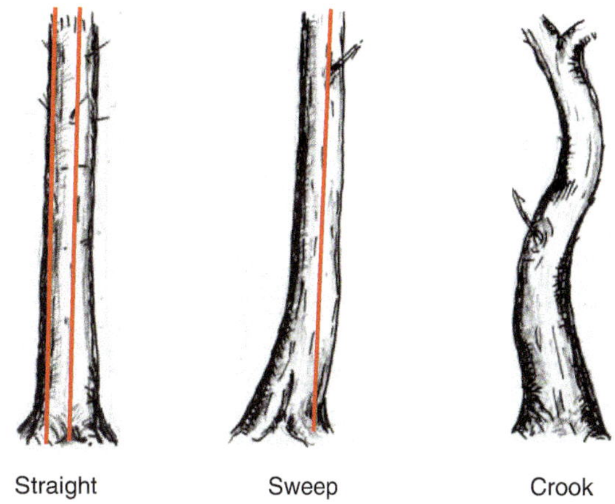

Fig. 5.10 Variations in stem crookedness

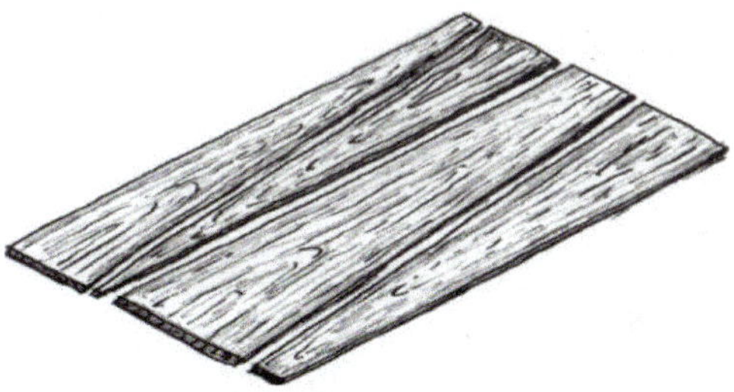

Fig. 5.9 Boards are cut from a tapered stem and trimmed parallel to the wane

Fig. 5.11 Sorting timber for shipbuilding from crooked oak trees (*Quercus*) in the early nineteenth century (In: Radkau and Schäfer 1987, author unknown)

expanses of forestland. Crooked stemmed trees accounted for 80 % of the felled timber with straight stems making up only 20 % (Fleischer 2009), (Fig. 5.11). At this time, the first grading standards started to emerge in an effort to establish rules for classifying wood quality based on intended end use.

Eventually, demand for crooked stemmed timber declined as sawmills capable of handling only straight-stemmed logs grew in popularity. Important developments in this process were the inventions of the water-powered sawmill by Villard de Honnecourt in the year 1230, crosscut saw by Stephanus Vinandus Pighius in the year 1587, buzz saw by L. A. C. Albert in the year 1799, rip saw by S. Bentham, around 1800 and band saw by Fa.Permin in the year 1852 (In: Finsterbusch and Thiele 1987).

By the late nineteenth century, specific guidelines for classifying wood quality were in place, setting, for example, the standard for maximum stem curvature permitted in a specific quality class. In Germany, the quality classes C and D expressly promoted "sawability" with a maximum allowable crookedness per lfd. meter (HKS 2002a).

Measurement: According to German timber grading standards, valid until 2012, stem crookedness is determined as follows: "stem crookedness is measured by dividing the distance between the two measurement points - expressed in centimeters and rounded to the nearest centimeter – by the distance that separates both ends of the curved stem in meters. The curvature is expressed in centimeter per meter" (HKS 2002a, p. 10).

Specific guidelines for measuring the length of the stem curvature, however, were not included. In practice, though, sweep and crook were not measured along the length of the curved stem axis, but always as the shortest distance between the stem ends.

According to the timber grading standards of the European Union (CEN), "Timber with crook is measured in separate sections. The timber should be separated into straight or sweep sections. Each section is separately measure as straight timber. The individual lengths are added together. Each section should have a minimum length that corresponds to the minimum length in the appropriate European standard, meaning that measurements are taken between the shortest distance of the turning point the curvatures (DIN 1998b, p. 4).

The European guideline refines the previous HKS practice by calculating the stem length as the sum of the lengths of straight or sweep sections. The *method for measuring*

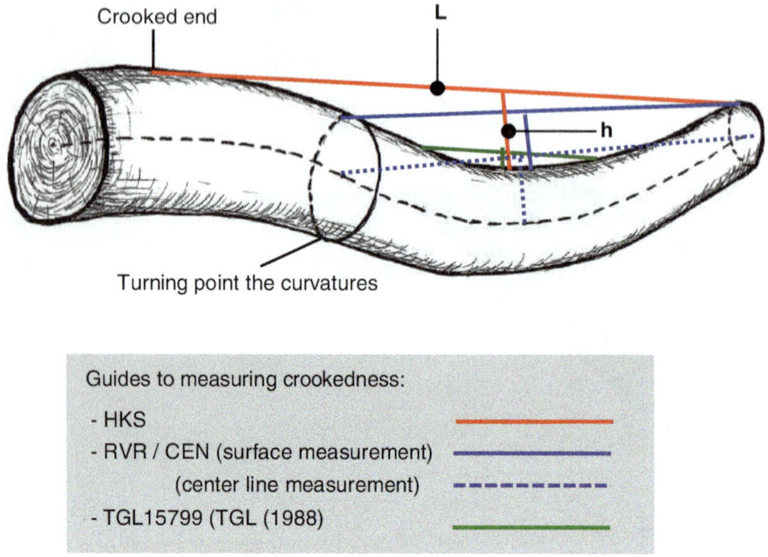

Fig. 5.12 Various types of crookedness measurements

dimensions is set in the grading standards: "When possible divide the maximum distance between the accepted concave centerline of the stem section and a line connecting the centers of both log ends by the length of the log, in a ratio of centimeters over meters." This applies to the optoelectronically measured and reproduced stem projections…. "If this is not possible then measure the maximum distance between the concave surface and a straight line connecting the innermost points of both log ends" (DIN 1998b, p. 11).

The "Framework Agreement for Timber Trade in Germany" (RVR 2014), valid in Germany since 2012, follows the European grading standards for stem crookedness (CEN) but requires measurements on butt logs to be taken a meter inwards from the stem base.

Although the grading rules are formulated differently, the measuring principle remains the same – measurement point h, the greatest distance from stem surface along the line adjoining both ends of the curvature (or the stem midline for optoelectronic measurements), is set in proportion to L, the length of this line (Fig. 5.12):

$$\text{Crookedness}\left[\text{cm}/\text{m}\right] = \frac{h\ [\text{cm}]}{L\ [\text{m}]}$$

In practice, measuring stem crookedness in the forest is difficult and inaccurate and, therefore, not generally recorded. In countries with modern wood processing technology, measurements for crookedness are generally made at the sawmill and are taken at several different points equally distributed along the length of the logs. In Germany, the standards are set by the *Framework Agreement for Sawmill Log Measuring* (Rahmenvereinbarung Werksvermessung 2005).

See Plate 5.2 for stem curvature photos.

Dwarf beech (*Fagus sylvatica* var. *tortuosa*) with genetically caused crookedness (Tharandt, Germany)

Birch (*Betula pubescens* var. *tortuosa*) pushed down each winter by snow (Finnmark, Norway)

This stand of crooked oak, maple, and ash (*Quercus* ssp., *Acer* ssp., *Fraxinus excelsior*),from a coppice grew without management (Erz Mountains, Germany)

A crooked oak (*Quercus* ssp.) straightens after decades of suppression under crown competition

Crooked growth pine (*Pinus sylvestris*) from unsuitable seed supply on loessic loam (agroforestry) on a private woodlot

Spruce (*Picea abies*) with sweep caused by soil creep on tertiary ground straighten themselves through photo- and geotropism (Kalkalpen, Germany)

Sable growth Douglas fir (*Pseudotsuga menziesii*) at 2,000 m elev. (Montana, USA)

Visible after logging in a tropical forest: second-story trees curve in an effort to reach the canopy (Surinam)

Plate 5.2 Stem crookedness

This crooked elm (*Ulmus glabra*) finds a buyer because of its rare stem shape (Germany)

This crooked pine (*Pinus sylvestris*) is cut at the point of greatest crookedness to produce a straight log

Crooked spruce (*Picea abies*) board with compression wood splits at the point of greatest pressure

Board from a curved poplar (*Populus tremula*) stem with reaction wood splits down in the area of the greatest tensile stress

Bush tamarind (*Mimosa guianensis*) is so valuable that even extremely crooked stem pieces are milled

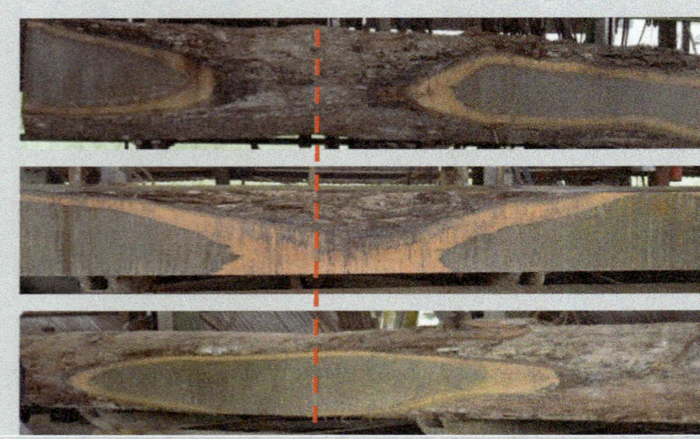

Yield losses of a valuable but crooked greenheart (*Tabebuia serratifolia*) stem due to a faulty crosscut. Three sides of the stem shown, each with significant amount of scrap wood (Surinam)

Significant amount of waste cut from a crooked beech (*Fagus sylvatica*) log

Pine (*Pinus sylvestris*) with marked wane and rough surface (flat cut)

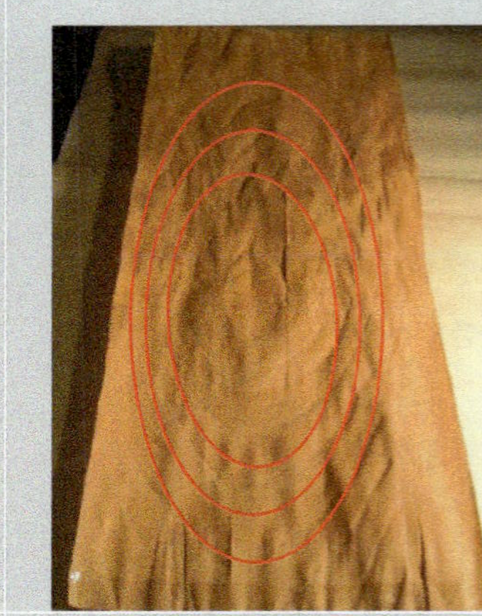

Veneer with elliptic growth rings, made from a crooked beech (*Fagus sylvatica*) log

Plate 5.2 (continued)

Timber-framed house built with naturally shaped struts and braces (Lausitz, Germany)

Wooden bridge from 1720 made with naturally shaped stem parts (Aarberg, Switzerland)

Woodshed with half timbers from a crooked stem (Lower Saxony, Germany)

Rudder bracket from naturally shaped walnut on a gondola (Venice, Italy)

Plate 5.2 (continued)

5.1.2.2 Causes

Certain tree species and their provenances have a genetic predisposition to crook, e.g., pine (*Pinus sylvestris*), or sweep, e.g., various larch provenances (*Larix* ssp.). In the nineteenth century, the severely twisted growth of the dwarf beech (*Fagus sylvatica* var. *tortuosa*), a mutation of the common beech (*Fagus sylvatica*), from the Süntel hills of Lower Saxony, caused a huge sensation.

Trees tilted from their vertical position by external forces (soil creep on hillsides, wind and snow pressure) straighten themselves as a result of heliotropism and geotropism. In doing so, the stem axis bends (Fig. 5.13). A tree's ability to bend its stem and branches through the formation of reaction wood is vital to its survival.

The tree must be able to orientate its terminal shoot against gravity and always in the direction of the most abundant supply of light (as exhibited by trees growing under heavy crown coverage or on the boarder of a stand with light only from one direction).

If a tree loses its terminal shoot, a side branch will take over the leader function ("post horn" or "bayonet growth" in softwoods). This type of damage is often caused by dear browsing or fraying, especially in young stands (Fig. 5.14).

Boring insects such as the pine shoot moth (*Rhyacionia buoliana*) cause deformed tree shapes (post horn).

Trees which grow out from the lower and secondary stories under crown competition are forced to orientate their main shoot towards a gap in the canopy with the most sunlight. The crooked stem can continue to grow, but will produce tension internally (Fig. 5.15). Particularly in the tropical primary and secondary forests, this internal defect in the fiber can have a significant impact on future processing (warping, cracking) (Harzmann 1988).

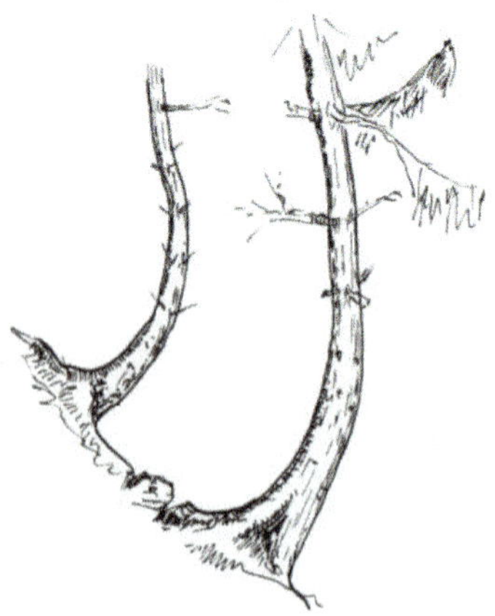

Fig. 5.13 Spruce (*Picea*) with sweep on a hillside (soil creep)

Fig. 5.14 Bayonet-shaped pine (*Pinus*) after crown break due to browsing or fraying by deer

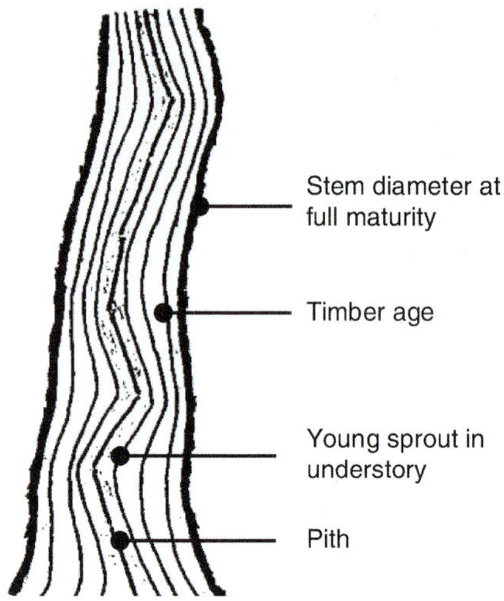

Fig. 5.15 Crooked stems can straighten as the tree grows

5.1.2.3 Prevention

Strict adherence to seed and stock selection guidelines during artificial stand formation can prevent providences with a tendency towards stem crookedness from passing on their genetic potential.

Spacing guidelines and selective thinning promote stand members predisposed to good growth and well-formed stems (Fig. 5.16).

Formation pruning of seedlings or immature hardwood stands in an effort to prevent future stem curvature is no longer commonly practiced because in most cases, despite pruning, the terminal shoot will eventually still differentiate itself from the other shoots. In softwood stands, trees with curved stems are usually removed because pruning can result in enclosed bark and decay.

Trees with crooked stems in younger stands, most notably oak, straighten themselves as the stand matures. Trees with sufficient crown space develop wide growth rings, particularly in the concave side of the curvature. This can considerably improve stem quality by the time of cutting maturity (Bues and Weiß 2002).

Fig. 5.16 Removing curved trees as part of stand management

Fig. 5.18 Scrap wood (*red*) from two planks milled from a crooked log

Fig. 5.17 "Waste wood" in peeled veneer from a crooked veneer roll

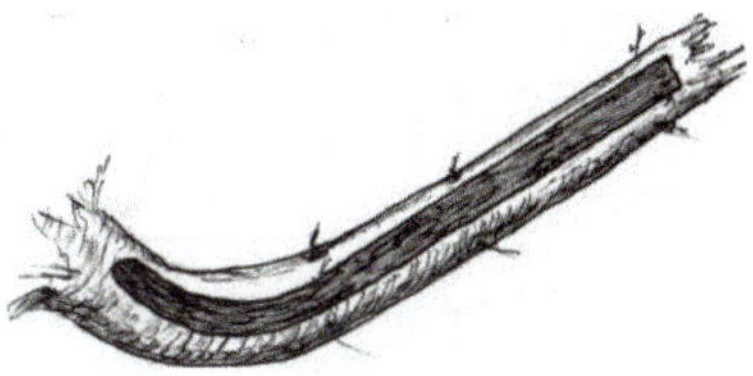

Fig. 5.19 Sled runner made from a crooked stem base (Lapland, Finland)

5.1.2.4 Impact on Use

In general, stem crookedness is always associated with the formation of reaction wood (compression wood in softwoods, tension wood in hardwoods). This has an unfavorable effect on processing and the dimensional stability of the wood products.

Crooked stem pieces used to produce figured veneer have a lower yield.

Using crooked wood to make peeled veneer produces a significant amount of waste, even though the peeling rollers are automatically centered on the wood (Fig. 5.17). The wood fibers are cut diagonally. This reduces the strength and dimensional stability of the veneer.

In the production of sawn timber, the large proportion of slab and end cuts reduces the volume yield (Fig. 5.18). The section of missing fiber reduces the strength of the wood and increases the surface roughness.

Paints and varnishes are absorbed differently, depending on the direction of the wood fiber.

If an object can be made by maintaining the natural curve of the wood, it will increase the strength of the end product (e.g., boat frame, sled runner, ice hockey stick) (Fig. 5.19).

Although it is sometimes possible through forest management practices to prevent stem crookedness, the many diverse biotic and abiotic impacts affecting trees are generally so significant it is rare to find a tree stem that has maintained perfectly rotationally symmetry throughout its often century-long life.

The effects of stem crookedness on the quality classes can be quantified according to the *Framework Agreement for Timber Trade in Germany* (RVR 2014) as follows (Table 5.3):

5.1.2.5 Technological Adaptation

Trees are purposely felled in a manner that allows logs with simple sweep to still be effectively milled. Usually, the cross-cut is made at the point where the stem begins to curve (Fig. 5.20).

Today, crooked stems can be processed using log breakdown systems with hydraulic controls that guide the logs through the gate or band saw (Fig. 5.21). However, as the board or plank strength increases, it becomes more difficult to avoid producing scrap wood. During drying, the wood pile must be fixed flat.

Following the principle "If you can't fix it, use it to your advantage," sawmills in Switzerland have developed new technology programmed to cut boards following the logs' natural contours. These curve saws not only limit waste but also create more homogeneous products with greater commercial value (Fig. 5.22).

Stem crookedness and stem taper make measuring wood volume difficult for industrial mills. To overcome this problem, processors rely on (semi)automated measurements. The measurements are never exact but close enough to provide a working basis.

Table 5.3 Effects of sweep on quality class according to RVR (RVR 2014)

Species	Average diameter	RVR quality class			
		A	B	C	D
Spruce/fir (*Picea abies/Abies alba*)	Ø<20 cm	≤1 cm/m	≤1 cm/m	≤1.5 cm/m	≤3 cm/m
	Ø≥20<35 cm		≤1 cm/m	≤1.5 cm/m	≤3.5 cm/m
	Ø≥35 cm		≤1.5 cm/m	≤2 cm/m	≤4.5 cm/m
Pine (*Pinus sylvestris*)		≤1 cm/m	≤1.5 cm/m	≤3 cm/m	≤4.5 cm/m
Douglas fir (*Pseudotsuga menziesii*)	Ø<35 cm	≤1 cm/m	≤1.5 cm/m	≤2.5 cm/m	≤4.5 cm/m
	Ø≥35 cm	≤1.5 cm/m	≤2 cm/m	≤3 cm/m	
Larch (*Larix decidua*)	Ø<35 cm	≤1 cm/m	≤1.5 cm/m	≤2.5 cm/m	≤4.5 cm/m
	Ø≥35 cm	≤1.5 cm/m	≤2 cm/m	≤3 cm/m	
Oak (*Quercus* ssp.)		≤2 cm/m	≤4 cm/m	≤10 cm/m	Unlimited
Beech (*Fagus sylvatica*)		≤2 cm/m	≤5 cm/m	≤10 cm/m	Unlimited

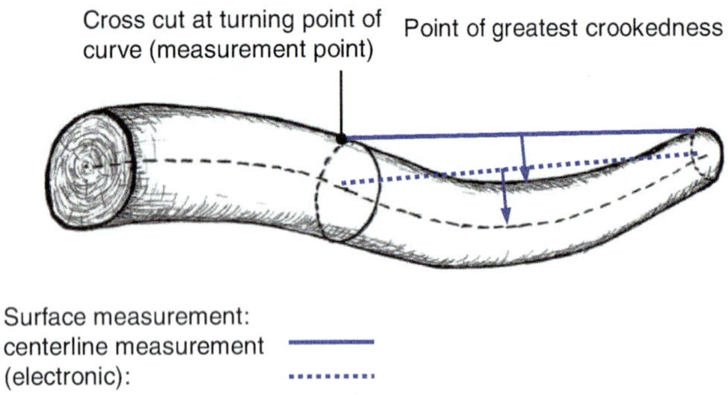

Fig. 5.20 Crosscut of heavily curved stem

Fig. 5.21 Hydraulically controlled cut through a curved stem

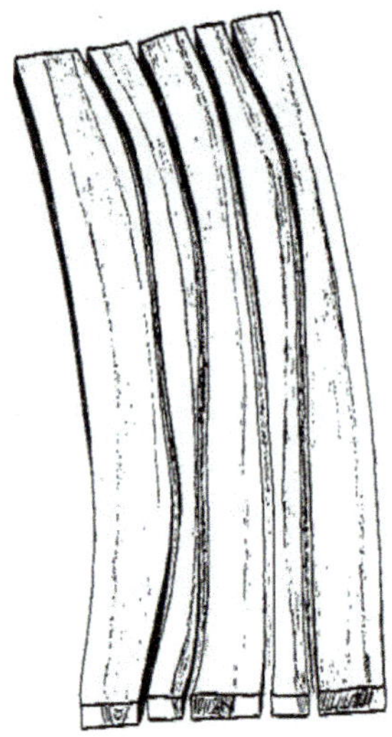

Fig. 5.22 Crooked walnut boards (*Juglans regia*) are cut with a curve saw into planks

5.1.3 Forking (Richter 2002b)

5.1.3.1 Description

Forked stem growth is grouped in two categories: false and true forks.

False forks, also called fused trunks, develop when two stems grow together at their base during radial growth and begin forming mutual growth rings. A true fused trunk is only possible between trees of the same species. Different species develop mutually deformed stems but no fused trunks.

Trees with several stems growing together are called multi-trunk trees (Fig. 5.23).

True forks, also called codominant stems or bifurcations, develop when at least two stems of similar dimension grow out of the same tree bole. Before the forking, the bole has only one pith (Fig. 5.24).

If the angle of the fork is wide enough to permit continued radial growth in the fork's crotch, the tensile force from the weight of the crown will be equally distributed between the stems thereby making them stronger. The result is a u-shaped crotch, also called a tension fork (Mattheck 1997).

True forks that develop into u-shaped crotches have long been considered a valuable building material, particularly in shipbuilding and farming. The fork's naturally reinforced fiber forms a strong union between each part of the crotch (Fig. 5.25).

When the fork angle is narrow (v-shaped crotch), there is danger that the stems' radial growth will progress faster than vertical growth in the crotch. In this case, the bark in the increasingly expanding intersection wedges together and becomes enclosed. As a result, the union between the stems is weak and prone to splitting. The v-shaped fork is also called a compression fork because the successive growth in the crotch area leads to an increase in compression force (Mattheck 1997).

As radial growth continues in the stems, swelling or compression folds, also called elephant ears or noses, form in the crotch area. The danger of breakage increases as the compressions folds become more prominent and stronger, even if they once again form mutual growth rings. The transition from tension fork to compression fork is illustrated in Fig. 5.26.

According to studies by Funke et al. (2011), bark enclosures in beech (*Fagus sylvatica*) begin with a narrowing of the bifurcated stem angle starting at approximately 30° (compare Fig. 5.24).

The length of a stem with a forked crown is measured at the point at which the pith separates.

See Plate 5.3 for forking photos.

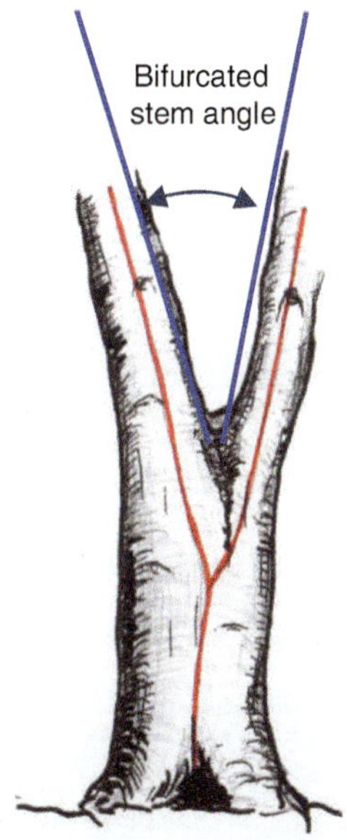

Fig. 5.24 True fork/codominant stems (*pith channel red*)

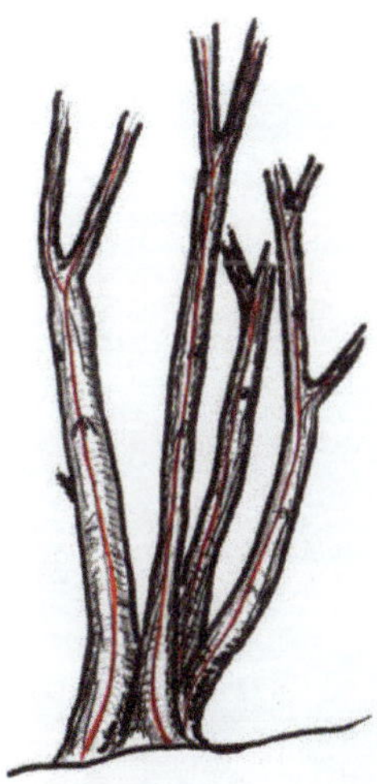

Fig. 5.23 False fork/fused trunk (multi-trunk stem)

Fig. 5.25 Naturally shaped forks, root, stem, and branches used in shipbuilding (replica of Viking ship (Roskilde, Denmark))

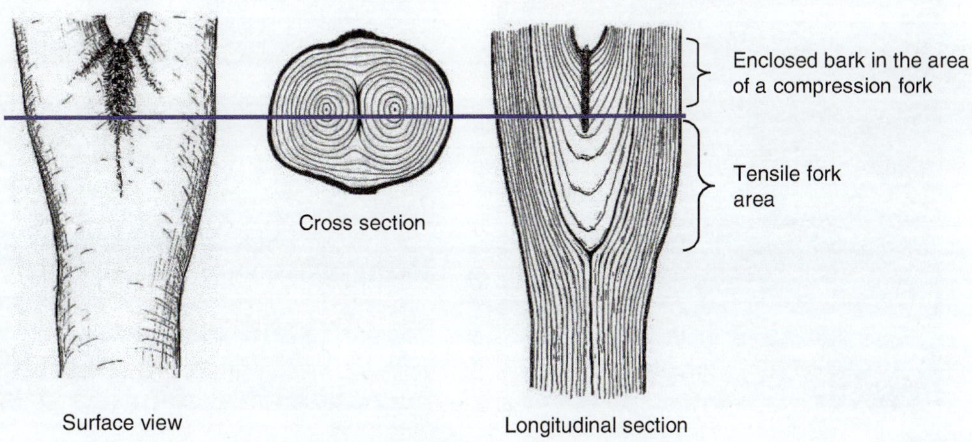

Fig. 5.26 Various views of a fork crotch

Broom-crowned, fork prone birch (*Betula pendula*) 🔴 and single-top birch crowns 🟢 on the same site

False fork between an oak (*Quercus robur*) and a beech (*Fagus sylvatica*)

Multi-trunk linden (*Tilia cordata*) after coppicing

False forking of two ash (*Fraxinus*) with "water pot" in a former coppice

False fork in a several-hundred-year old redwood (*Sequoia sempervirens*) (California, USA)

Acute fork (compression fork) in an old beech (*Fagus sylvatica*) with swelling (elephant ears)

Broken compression fork in a beech (*Fagus sylvatica*)with visible enclosed bark

Cut-through beech (*Fagus sylvatica*) wood *left* with enclosed bark, *right* tensile forking (continuous tree ring pattern) in lower and compression forking ⭕ in upper section

Plate 5.3 Forking

Tensile forking shaped like a bayonet in a pine (*Pinus sylvestris*) caused by snow damage to the crown

Ash bud moth (*Prays fraxinella*) damage in the terminal shoot leading to forking

Fork due to fraying damage by a rein deer on the terminal shoot in a pine (*Pinus* ssp.) (Lapland, Finland)

Split beech (*Fagus sylvatica*) fork on a forest path presents a hazard to pedestrians

Split fork in a valuable checker tree (*Sorbus torminalis*)

Oak (*Quercus robur*) branches grow together relieving the compression fork

Protective piece in veneer cheery (*Prunus avium*). Cross section in fork base with enclosed bark

Beech (*Fagus sylvatica*) damaged in logging. The fork split the entire stem in half

Forked oak (*Quercus* ssp.) stem used as construction wood in a 200-year-old block house (Bavaria, Germany)

Plate 5.3 (continued)

Aesthetic effect of reverse forking in cedars (*Cedrus*) (Chenonceau, France)

Trees in tropical primary forests rarely form forks able to survive, Swietbonki (*Inga* ssp.) (Surinam)

Forked in growth on a pine (*Pinus Ponderosa*), dead for decades (Utah, USA)

Trees growing in the tropics without crown competition also form forks like this mahogany (*Swietenia macrophylla*) (Surinam)

The stem surface already enclosed the fork in a vochysia (*Vochysia tomentosa*) (Surinam)

Extreme tensile fork in a eucalyptus tree (*Eucalyptus* sp.) (Madeira, Portugal)

Pyramid veneer from a toppled eucalyptus (*Eucalyptus* sp.) (From Danzer, Furniermagazin 2008, p 39)

Pyramid veneer made from mahogany (*Swietenia macrophylla*).

Plate 5.3 (continued)

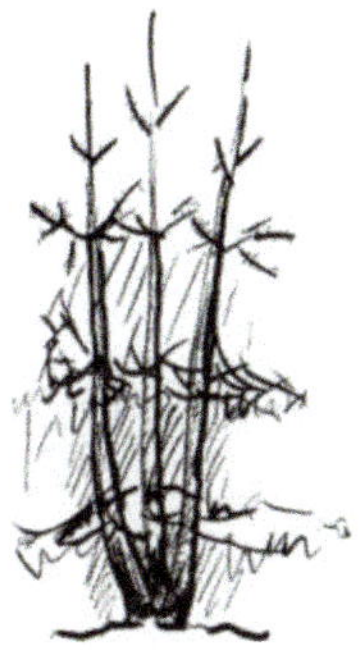

Fig. 5.27 Spruce *(Picea abies)* bunch planting

Fig. 5.28 Genotypic forking in a birch (*Betula pendula*), right single-top birch crown

5.1.3.2 Causes

False forks are often caused by bunch planting, thinning (coppice system), or unmaintained natural regeneration. Germinating seeds from a squirrel's food stash can also lead to false forking (Fig. 5.27).

True forks are usually genotypically induced. As a result, certain types of hardwoods (birch (*Betula pendula*), beech (*Fagus sylvatica*)) are prone to forking (Fig. 5.28).

If a terminal bud or young terminal shoot is destroyed, several side branches will generally take over the function of the terminal shoot. Insects cause damage by boring into the terminal bud (ash bud moth (*Prays fraxinella*), pine shoot moth (*Rhyacionia buoliana*)), but also birds use the freshly budded leaves from shoots as nesting material to improve brooding conditions. Terminal shoots are also broken by snow and ice.

Browsing and fraying damage from antlered game leads to low forking. Tree species with opposite budding tend to grow side shoots that become forks. Softwoods and hardwoods such as maple (*Acer*), ash (*Fraxinus*), and horse chestnut (*Aesculus*), therefore, exhibit a strong tendency to fork, while tree species with alternate budding, such as oak (*Quercus*), beech (*Fagus*), and linden (*Tilia*), are less prone.

Given the heavy sociological competition in tropical primary and secondary forests, forking is only possible in the lower sections of the stem. As solitary trees, however, tropical species tend to fork just as hardwoods do in temperate zones.

Fig. 5.29 Removal (belated) of low-set forking as part of a beech (*Fagus sylvatica*) crown thinning

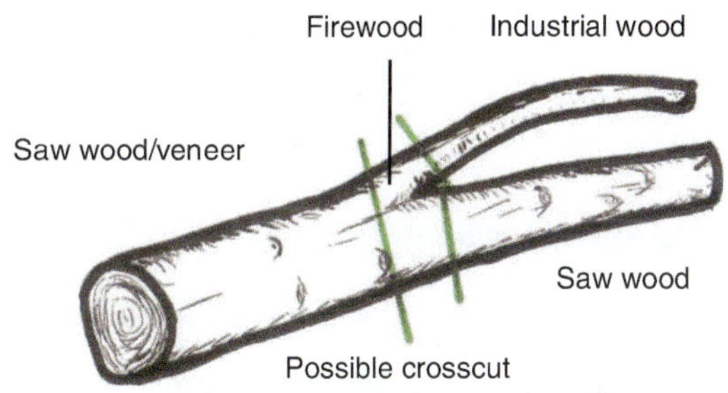

Fig. 5.30 Possible yield from a forked log

5.1.3.3 Prevention

Site-appropriate and well-formed tree species or provenances should be selected when forming a new stand.

Forking is best prevented through timely maintenance and effective thinning (negative selection) of immature stands (Fig. 5.29).

Forest protection measures are recommended to protect against browsing and insect damage (see Sect. 6.2.1).

Pruning young hardwoods is usually not worthwhile because the lead shoot will still eventually differentiate. However, for opposite branching species (maple (*Acer*), ash (*Fraxinus*)), pruning can be effective.

5.1.3.4 Impact on Use

False forks are wood defects, because they significantly affect the use of the wood. They are only considered not to be defects if they are intentionally kept to protect a log from splitting and not included as part of the log's length measurement. False forks usually require a scaling deduction.

The risk of breaking during storms and under snow pressure is higher for forked stems and they are prone to splinter and crack during logging.

Forked timber has shorter effective length for high-quality sawing or veneer wood and an increased share of industrial wood/plywood and lumber of inferior quality (Fig. 5.30).

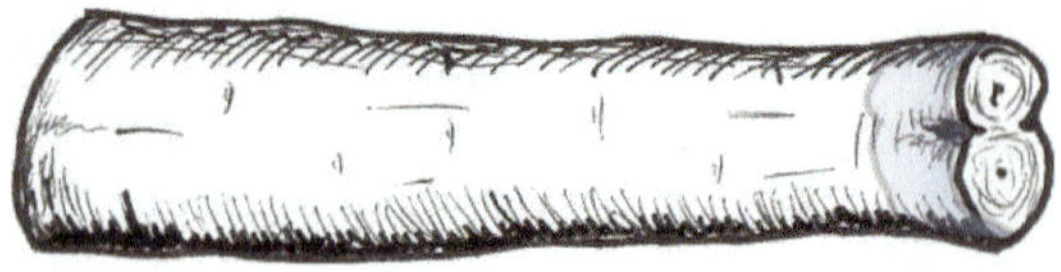

Fig. 5.31 Fork base left for protection on a veneer log

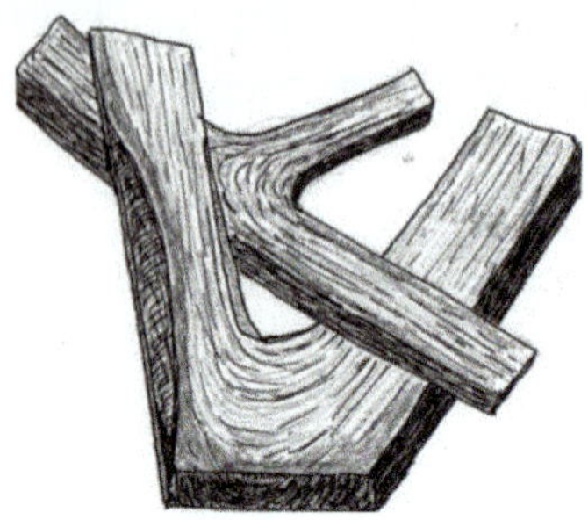

Fig. 5.32 Ship frame from a replica of a Viking ship, Roskilde, Denmark

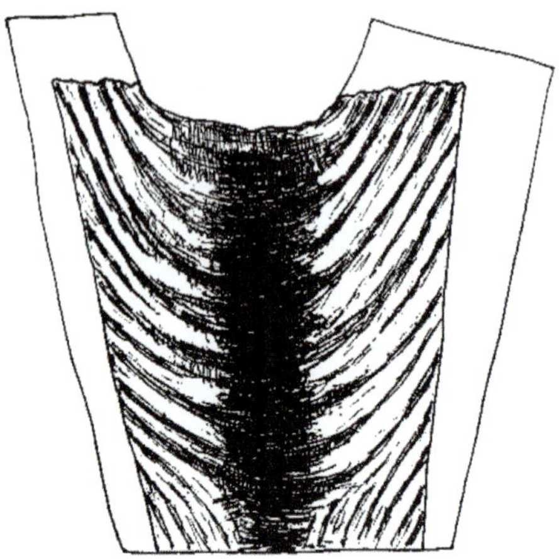

Fig. 5.33 Location of pyramid inlay from rosewood (*Dalbergia* sp.) in a fork

Enclosed bark or "water pots" in the crotch devaluate the wood (decay). Small forked stems are often unround and crooked at their basis and form reaction wood.

5.1.3.5 Technological Adaptation

Higher valued logs are crosscut through the fork base. The fork base is kept until further processing to protect the end of the stem from splintering (Fig. 5.31).

Forks have traditionally been used as building material, especially in shipbuilding as natural junctions and connecting elements (frames, sternpost) and for farming equipment (forks, mounts, grips, tool handles) (Fig. 5.32).

Today, skilled woodworkers use forked wood to create artistically sophisticated objects.

Despite the considerable disadvantages, forks are used in a few special applications. A strong, well-formed fork can be cut to create interesting wood figure, as, for example, the coveted pyramid texture veneer (Hoadley 1990), (Fig. 5.33).

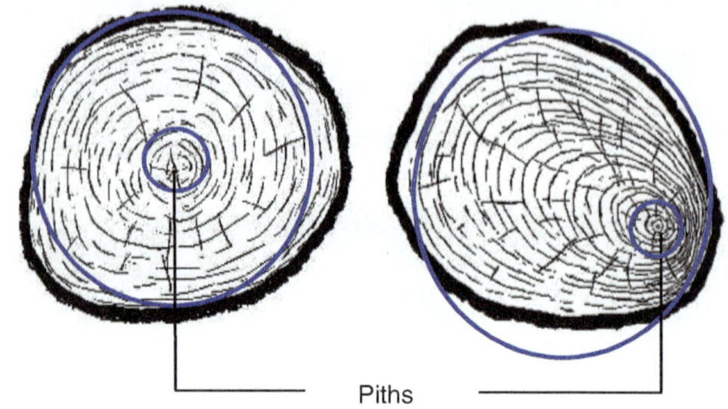

Fig. 5.34 Oval stem cross section with centered (*left*) and eccentric (*right*) piths

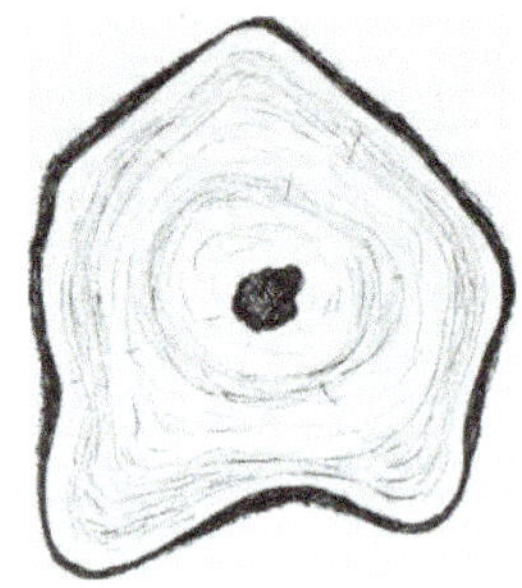

Fig. 5.35 Cross section from a kopi stem (*Goupia glabra*) with heart rot and tension wood folds

Noteworthy are the South American Swietenia mahogany (*Swietenia macrophylla*) and African *Khaya* mahogany (*Khaya ivorensis*). Rare forked growths of more than 2 m can occur in the latter (Furniermagazin 2003).

5.1.4 Out-of-Roundness

5.1.4.1 Ovality/Eccentric Growth (Richter 2002c)

5.1.4.1.1 Description

Ovality refers to a tree stem cross section with a significantly noncircular shape. The pith, however, is still located in the center of the cross-sectional disc (Fig. 5.34, left).

In stems with *eccentric growth*, on the other hand, the pith is located off-center in the cross-sectional disc (Fig. 5.34, right).

An oval cross section does not necessarily have an off-centered pith. Conversely, a cross section with an off-centered pith does not necessarily have an oval shape. Ovality and eccentric growth are always associated with variances in growth ring width and usually with reaction wood, particularly compression wood in softwoods. Especially in hardwoods from tropical primary and secondary forests, pronounced tension wood folds often lead to misshaped cross sections (Fig. 5.35).

See Plate 5.4 for ovality/eccentric growth photos.

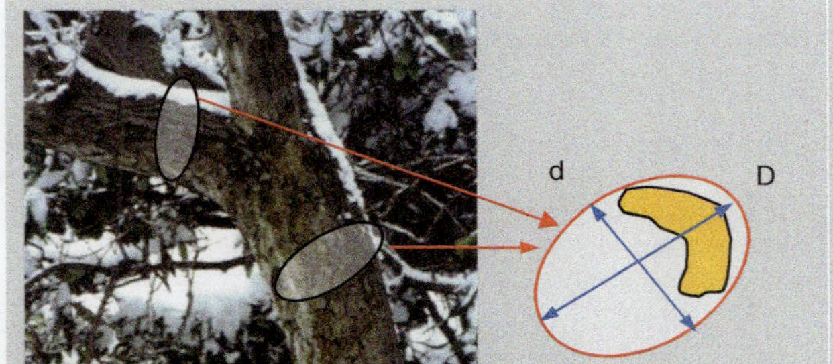

The D : d–ratio in the stem cross section is 7 : 5

Eccentric growth in a maple (*Acer rufinerve*) stem. Decades of lean resulted in tension wood formation (orange) (Tharandt, Germany)

Lawson's cypress (*Chamaecyparis lawsoniana*) branch righted itself by forming compression wood, leading to an eccentric cross section

Larch (*Larix decidua*) with highly eccentric stem base contour (saber growth)

Eccentric growth in a pine (*Pinus sylvestris*) stem with compression wood zones

Oval, circular, and eccentric cross sections from oak (*Quercus* ssp.) logs in a tender sale (Tharandt, Germany)

Slightly oval oak (*Quercus petraea*) stem

Extremely oval oak (*Quercus* ssp.) stem

Oval ash (*Fraxinus excelsior*) stem caused by tension wood

Plate 5.4 Ovality, eccentric growth

Kopi (*Goupia glabra*) with eccentric stem contour caused by tension wood folds (Surinam)

Gronfolo (*Qualea rosea*) with eccentric stem contour caused by extreme tension wood sections in the lower bole (Surinam)

Eccentric ingipipa (*Couratari fagifolia*) stems with tension wood folds (Surinam)

Extremely unround kopi (*Goupia glabra*) stem (Surinam)

Extreme out-of-roundness. Square cross section from the base of a tropical tree (Surinam)

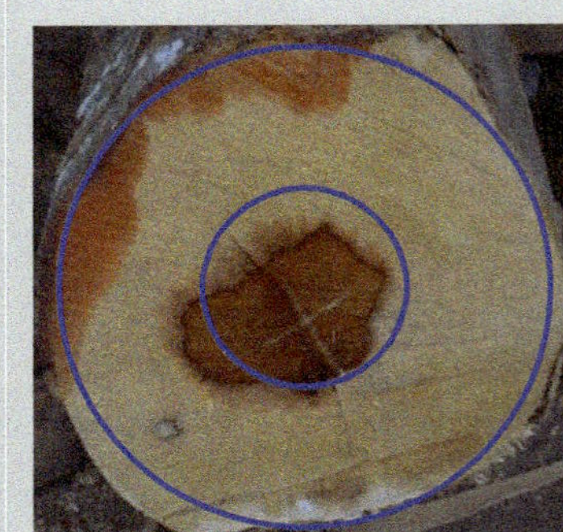

Unround baboon wood (*Virola surinamensis*) veneer roll with large portion of scrap wood (Surinam)

Layers of baboon *wood* (*Virola surinamensis*) (*light*) and walaba (*Eperua* sp.) (brown) before being manufactured into veneer (Surinam)

Plate 5.4 (continued)

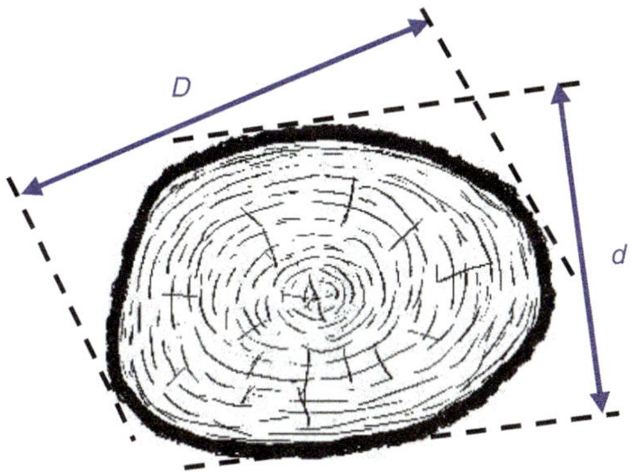

Fig. 5.36 Measuring ovality

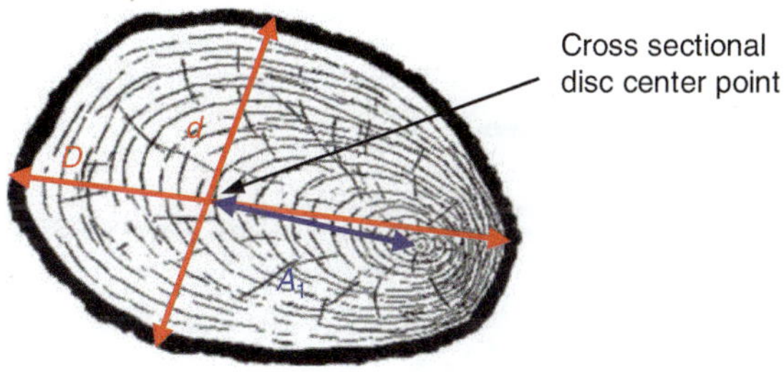

Fig. 5.37 Measuring eccentric piths (1)

Measurement: *Ovality* is measured as the difference between the greatest diameter (*D*) and the smallest diameter (*d*) (SSLEF 1997), or as the relationship between the smallest and the greatest diameter (Thüringer Landesforstdirektion 1997), or as the ratio between diameter difference and the greatest diameter, expressed in percent (DIN 1997b, 1998b) (equations and Fig. 5.36):

$$\text{Ovality} = D - d \; [\text{cm}]$$

or

$$\text{Ovality} = d : D \; (d = 1)$$

or

$$\text{Ovality} = \frac{D-d}{D} \times 100 \; [\%]$$

There are two methods for measuring eccentricity:

1. Eccentric growth is measured as the deviation of the pith (A_1) from the point of intersection between the greatest diameter (*D*) and smallest diameter (*d*) of the cross section of the butt or short wood. A_1 is related to the averaged diameter ($\phi = \frac{D+d}{2}$) of the cross section, expressed in percent (DIN 2002a) (Fig. 5.37):

$$\text{Eccentricity} = \frac{A_1}{\phi} \times 100 \; [\%]$$

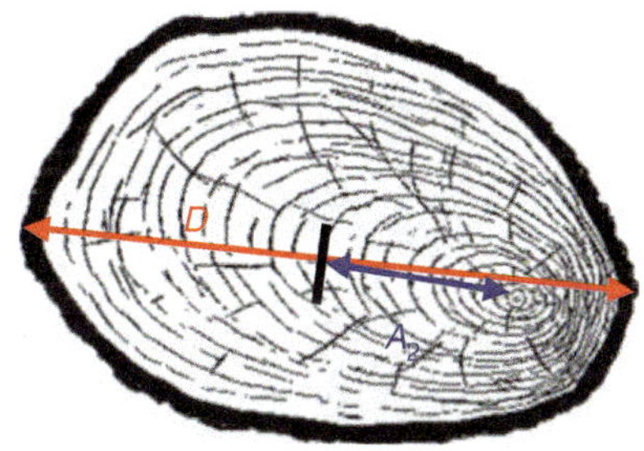

Fig. 5.38 Measuring eccentric piths (2)

Here, the reference point for the pith deviation is the center point of the cross-sectional disc. The CEN and RVR follow this approach whereby the *geometric center* serves as the reference point for the pith (RVR 2014). However, this method calculates somewhat higher values for pith deviation than the following method.

2. Eccentricity can also be measured as the deviation of the pith (A_2) from the center of the greatest stem diameter (*D*) on the stronger end of the stem. A_2 is related to the greatest stem diameter (*D*), expressed in percent (Frommhold 2001). Here, the reference point for the pith deviation is the center point of line *D* (Fig. 5.38):

$$\text{Eccentricity} = \frac{A_2}{D} \times 100 \; [\%]$$

When measuring the trunk base, it is advisable to eliminate the roots when calculating the center of the cross (Weidner 2000). This pragmatic advice is standard practice for measuring stump cross sections in many countries outside Europe.

In Germany, measurements follow the *Framework Agreement for Sawmill Log Measuring* (Framework Agreement for Sawmill Log Measuring 2005, p. 4). Accordingly, the measurements include "... *several evenly spaced measuring points, the number of which depends on the size of the log*". The oval butt ends are measured using other measuring points as comparison.

5.1.4.1.2 Causes

Oval stem cross sections with centered piths are likely to occur if a tree is crowed by adjacent tree crowns over a longer period of time. The reduced foliage in the affected crown area produces less assimilates, thereby restricting radial growth in the lower part of the stem.

On hillsides, stand borders, and steep (selection gaps), a constantly one-sided supply of sunlight (heliotropism) results in asymmetric crowns. Prevailing wind pressure or snow load can bring trees out of their vertical position. In such cases, hardwoods form tension wood and softwoods form compression wood in an effort to counterbalance the compression or tensile stress (Fig. 5.39).

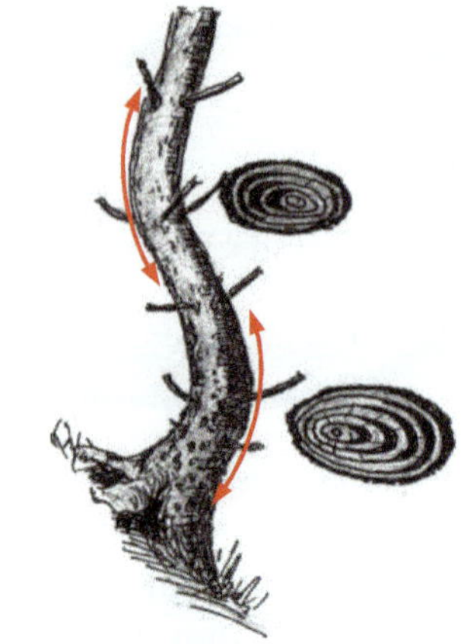

Fig. 5.39 Eccentric cross section in a spruce (*Picea abies*), stem that rightened itself with compression wood

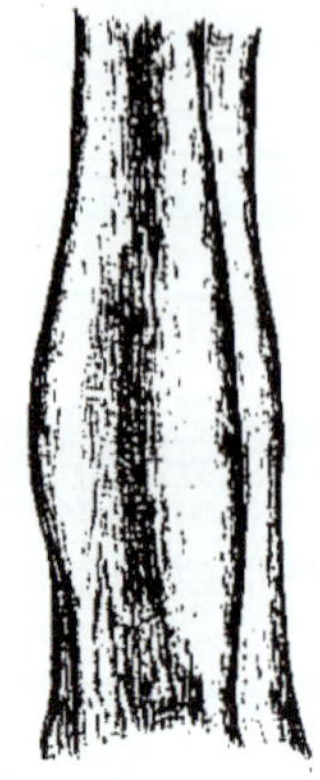

Fig. 5.41 Tension wood ridges in a kopi

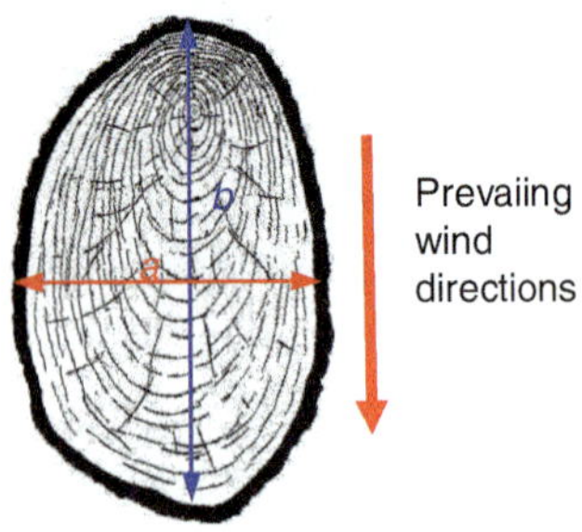

Fig. 5.40 Cross section of a pine (*Pinus sylvestris*) with eccentric pith

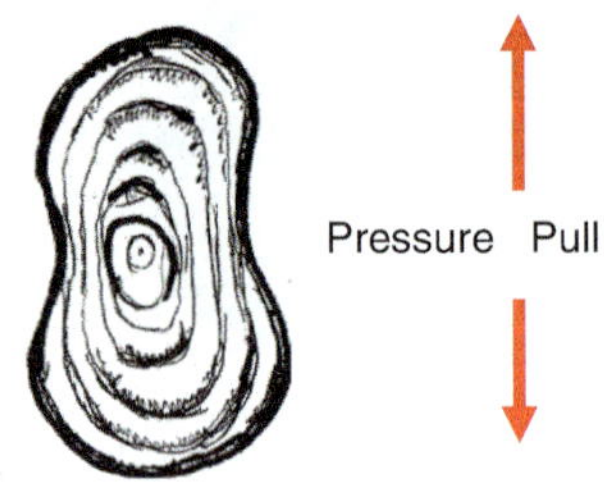

Fig. 5.42 Cross section of a spruce root near the root collar

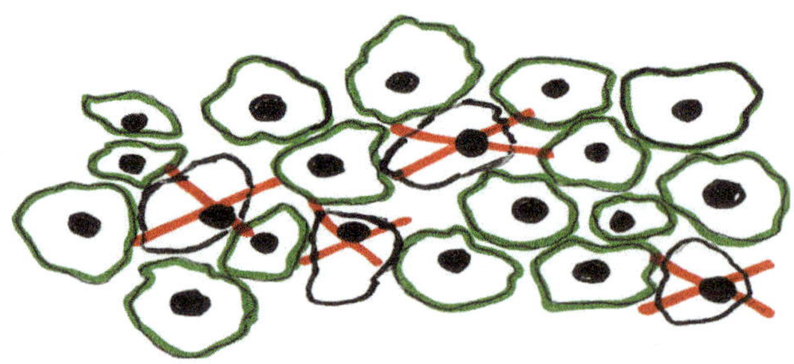

Fig. 5.43 Spacing in a beech stand to promote even crown development

Enlarged diameters are observed in many tree species in the prevailing wind direction (Fig. 5.40) (Mette 1984). A tree reacts to stress coming from a prevailing direction by acquiring a more or less eccentric stem shape. The moment of inertia area increases in the elliptic cross section as the greater diameter (*b*) exceeds the smaller diameter (*a*), because b to the third power is added to the calculation (Mattheck 1997):

$$\text{Moment of inertia area} = \frac{\pi}{4} \times a \times b^3$$

Different tree species are more or less prone to eccentric growth. The tendency towards eccentric growth declines in tree species from the temperate regions as follows (Grundner 1882): Pine (*Pinus sylvestris*) > larch (*Larix decidua*) > oak (*Quercus* ssp.)
>Beech (*Fagus sylvatica*) > spruce (*Picea abies*)

According to Mette and Boss (1964), as the stem diameter increases, so does the absolute diameter deviation.

Tree species growing in tropical primary and secondary forests need to be able to respond quickly to changes in the canopy. They form folds of tension wood that lead to extremely noncircular cross sections (Fig. 5.41; see Fig. 5.35). The out-of-roundness is often hidden once the tree assumes a dominate position in the canopy but revealed when the tree is logged (Harzmann 1988).

Unilateral stress particularly affects branches and root wood. This is due to their specific physical demands (Fig. 5.42).

5.1.4.1.3 Prevention

Ovality can generally be prevented through suitable spacing and regular crown maintenance over a stand's lifespan (Fig. 5.43).

To eliminate eccentric growth as much as possible, good stand management requires that a tree's main crown section

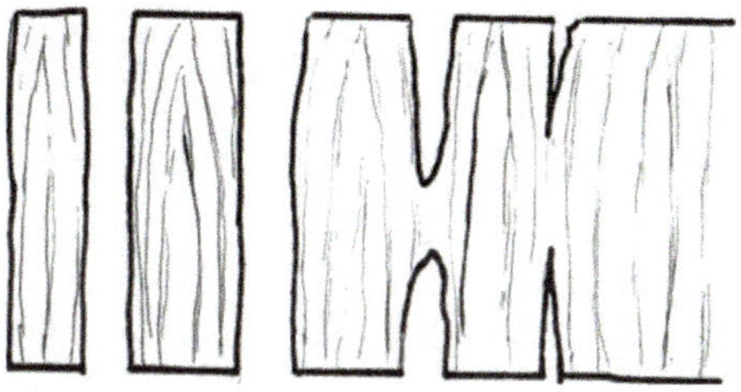

Fig. 5.44 Peel with varied figure

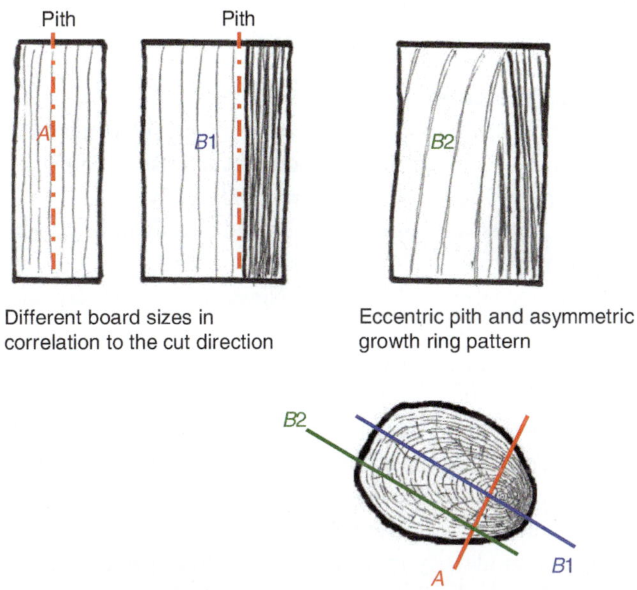

Fig. 5.45 Impact of cut on the growth ring pattern and sawn timber dimensions *A*, *B1* and *B2* = different cutting planes

lies within the center of the stem. Protection and appropriate spacing for the stand is also necessary.

In the tropics, growth dynamics in primary and secondary forests often lead to eccentric tree shapes. Tree plantations can limit out-of-roundness through systematic planting.

5.1.4.1.4 Impact on Use

"Peelers" during rotary cutting are associated with different degrees of figure, lower yields, and veneer quality (Fig. 5.44). Stems with sliced veneer quality are degraded because of the extreme pith eccentricity.

Eccentric pith in sawn timber results in varied growth ring widths. The actual board widths deviate from the grade of the cut stem (Fig. 5.45).

Because eccentric piths commonly correlate with reaction wood formation, they are also associated with variances in the swelling and shrinkage properties of the sawn timber.

Table 5.4 Effect of eccentric pith on quality class according to RVR (2012)

		RVR quality class			
Species		A	B	C	D
Spruce/fir (*Picea abies*/*Abies alba*)	Eccentricity of pith (%) (ovality is not shown)		≤15		
Pine (*Pinus sylvestris*)					
Douglas fir (*Pseudotsuga menziesii*)		≤10	≤20	Unlimited	Unlimited
Larch (*Larix decidua*)					
Oak (*Quercus* ssp.)					
Beech (*Fagus sylvatica*)					

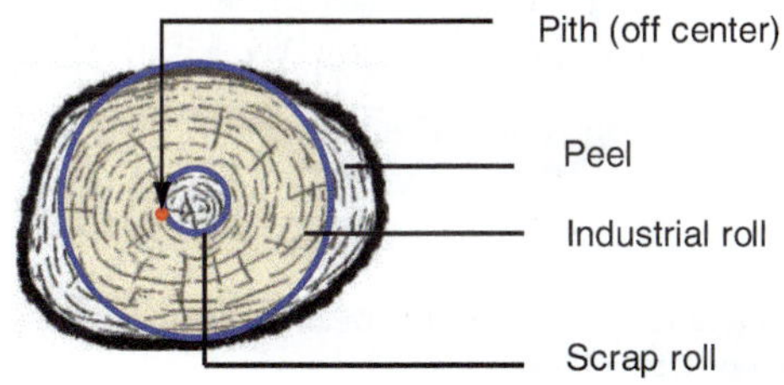

Fig. 5.46 Optimal positioning of an oval veneer roll

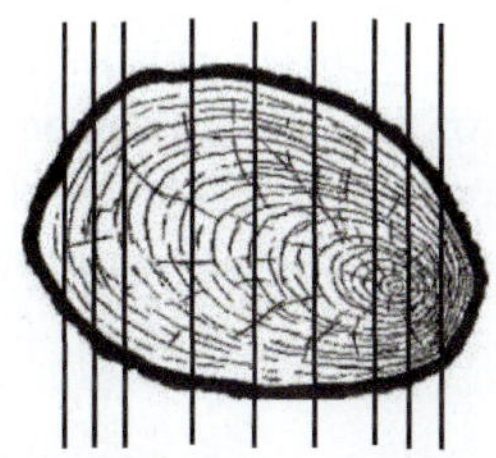

Fig. 5.47 Round cut, flat edge, with eccentric pith

The "Framework Agreement for Timber Trade in Germany" (RVR 2014) does not measure ovality. An eccentric pith is taken into account (Table 5.4).

5.1.4.1.5 Technological Adaptation

Rotary cut veneer processing automatically centers the cutting jig at the center of the cross-sectional face veneer sheet in order to achieve a maximum yield (Fig. 5.46). Meanwhile, computerized XY alignment using laser beam is common.

Stems with eccentric piths are cut by guiding the flat edge with the horizontally greater diameter through the saw so that boards can be produced with symmetrical growth rings (Fig. 5.47). In the following model cut, the

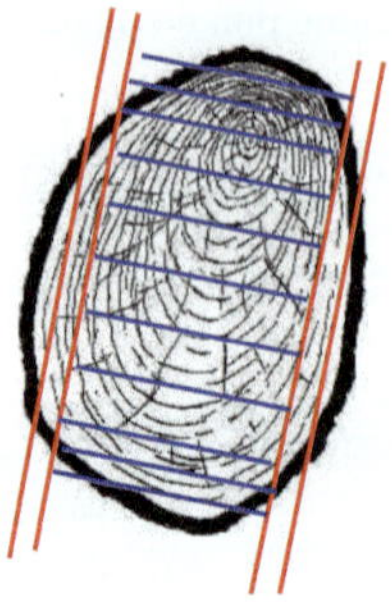

Fig. 5.48 Preliminary cut, vertical with following model cut in a stem with eccentric pith

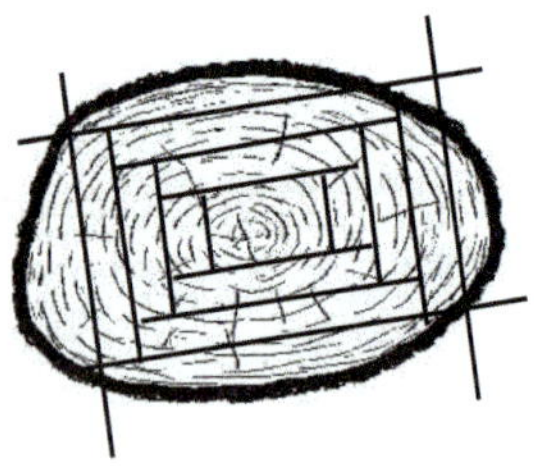

Fig. 5.49 Preferred cutting direction of an oval stem with centered pith using band saw technology

log is guided upright through the saw (Lohmann 2005) (Fig. 5.48).

In band saw technology, rotating blades are able to cut boards with symmetrical growth rings, if attention is given to ensure that the pith in the increasingly thinner model is centrically positioned and remains in the (residual) squared timber (Fig. 5.49).

Fig. 5.50 Beech (*Fagus sylvatica*) stem cross section above (*1*) and below (*2*) a moulding

5.1.4.2 Mouldings, Flutes, and Flanges (Richter 2005a)

5.1.4.2.1 Description

Mouldings, flutes, and flanges belong, along with ovality and eccentricity, to the group of wood characteristics known as stem contour irregularities (Frommhold 2001).

They are usually either not included or only briefly mentioned in international timber grading standards (HKS 2002a; DIN 1997c, 1998b; RVR 2014). Nevertheless, they are particularly interesting because their specific causes and impact on use vary significantly.

Mouldings are channel-like depressions or grooves running with the fiber along the length of the stem (SSLEF 1997). They generally start below a suppressed branch (shade branch) and usually continue down to the base of the tree. Mouldings associated with shade branches are particularly easy to identify in smooth-barked tree species such as beech (*Fagus sylvatica*) (Fig. 5.50), birch (*Betula pendula*), and yew (*Taxus baccata*).

See Plate 5.5 for images of mouldings.

Deep mouldings under suppressed branches on a 150-year-old yew (*Taxus baccata*)

Mouldings under suppressed branches in a pine (*Pinus sylvestris*)

Mouldings under a branch on a dawn redwood (*Metasequoia glyptostroboides*)

Moulding under a suppressed branch on an ancient beach (*Fagus sylvatica*)

Rare shade branch above a moulding (*Manilkara bidentata*) in a primary forest (Surinam)

Deep mouldings caused by extreme *left* spiral in the grain of this 200-year-old beech (*Fagus sylvatica*) (Rhön, Germany)

Deep moulding with enclosed bark at the base of a beach stem (*Fagus sylvatica*)

Valuable mountain elm (*Ulmus glabra*) included in timber sale despite moulding

The branches from zwart parelhout (*Aspidosperma excelsum*) only grow in the moulding and not on the statically important stem bole – perfect example of optimized stability and material economy (Surinam)

Plate 5.5 Mouldings

Flutes refer to deep, wave-shaped grooves in the stem surface running parallel to the stem axis (SSLEF 1997). The grooves can narrow to thin pleats or folds. They usually include an enclosed bark near the stem base, especially in hornbeam (*Carpinus betulus*) (Fig. 5.51 [2]), beech (*Fagus sylvatica*), birch (*Betula pendula*), juniper (*Juniperus communis*), yew (*Taxus baccata*), and cypress (*Taxodium*) and in tropical species such as B. witte parelhout (*Aspidosperma marcgrafianum*) (Fig. 5.55).

Hornbeams (*Carpinus betulus*) often exhibit structured net-like swelling on their stem surface. The growth rings are asymmetric (Fig. 5.51 [1]).

See Plate 5.6 for images of fluting.

Flanges are buttress-like formations protruding from the stem base. They create mouldings and folds in the stem surface extending several meters up the stem. They are particularly common in elms (*Ulmus*) (Fig. 5.52 [1]), cypress (*Taxodium*), and birch (*Betula pendula*), as well as many tropical tree species, as, for example, witte pinto locus (*Martiusia parvifolia*) (Fig. 5.52 [2]).

In tropical primary and secondary forests, moulding are rarely found under shade branches because these branches quickly die off under the canopy. Flutes and flanges, however, are very common and often found in the form of buttress roots.

See Plate 5.7 for images of flanges/buttress roots.

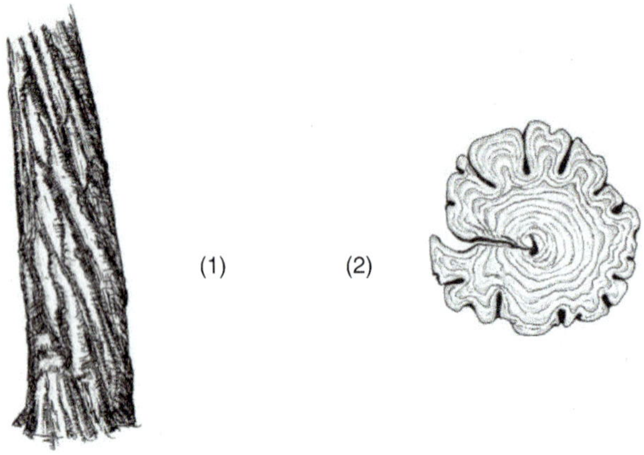

Fig. 5.51 Fluted stem base (*1*) and stem cross section of a horn beam (*Carpinus betulus*) with ingrown bark (*2*)

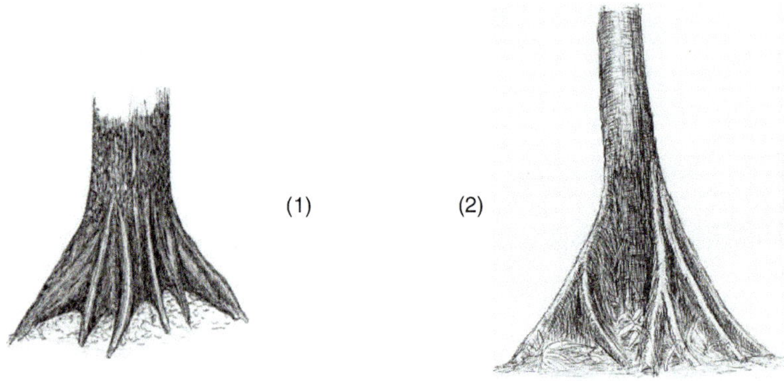

Fig. 5.52 Elm (*Ulmus*) (*1*) and witte pinto locus (*Martiusia parvifolia*) (*2*) with buttress-like flanges

Hornbeam (*Carpinus betulus*) with extreme fluting

Hornbeam logs (*Carpinus betulus*) at a timber sale

In birch (*Betula pendula*), flutes and flanges sometimes overlap

Board made from fluted fukadi wood (*Buchenavia* ssp.)

Witte parelhout (*Aspidosperma marcgrafianum*) stem consisting of intergrown wood segments, *right*, cross section of a young stem

Witte parelhout (*Aspidosperma marcgrafianum*) is valued mainly as construction wood but also makes beautiful furniture

Plate 5.6 Fluting

Mountain elm (*Ulmus laevis*) with buttress roots on a wet site. Right cutting a tree with buttress roots low keeps the stem from splitting (Oberlausitz, Germany)

Spruce (*Picea abies*) with flanges on a windy, high-altitude site (Thuringia Forest, Germany)

250-year-old fir (*Abies alba*) with flanges (Black Forest, Germany)

Wood marker made from the roots of a spruce (*Picea abies*) (South Tyrol, Italy)

Base of a giant sequoia tree (*Sequoiadendron giganteum*) (California, USA)

Ingipipa (*Couratari guianensis*) with buttress roots (Surinam)

A narrow stem is supported by broad, thin buttress roots (species unknown) (Surinam)

Witte pinto locus (*Martiusia parvifolia*) with extreme buttress roots. The increment zones run through both the stem (*upper right*) and buttress roots (*lower right*) (Surinam)

Plate 5.7 Flanges/buttress roots

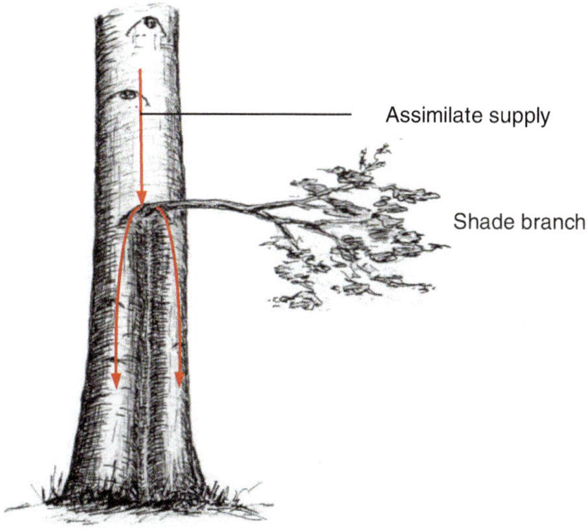

Fig. 5.53 Suppressed or shade branch in beech (*Fagus sylvatica*) creates a moulding beneath it

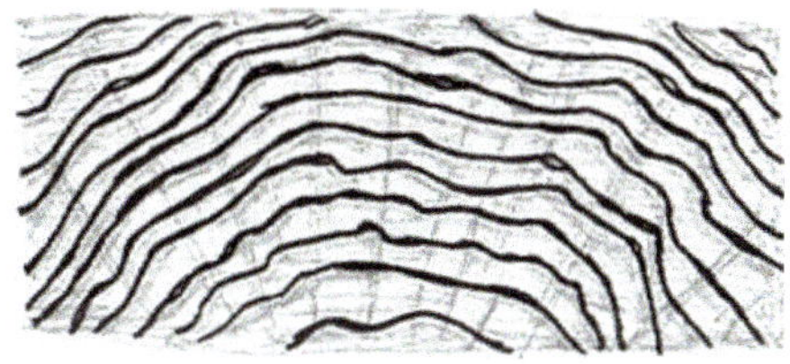

Fig. 5.54 Asymmetric growth rings in a fluted hornbeam (*Carpinus betulus*)

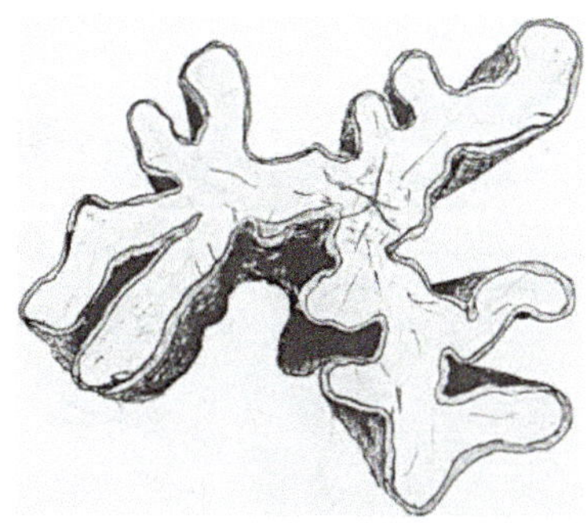

Fig. 5.55 Extreme fluting in a witte parelhout (*Aspidosperma marcgrafianum*)

Measurement (Mette 1984): A *moulding* is measured by determining the depth of the groove on the point of the greatest deviation from the stem's normal circumference.

Flutes are measured by determining the deepest grove in the radial direction, because this has the greatest impact on wood recovery.

Pronounced buttress-like *flanges* need to be removed before processing so they are not measured.

5.1.4.2.2 Causes

Although flutes and flanges often look like a series of moulding, they actually have very different causes and far-reaching implications on the wood's potential use.

Mouldings are frequently caused by a shortage in growth stimulants below a shaded, suppressed branch. The branches, so-called shade branches, use up the assimilates for themselves thereby depriving the xylem in the stem below (Grüner and Metzler 2003; Rubner 1910). The assimilate supply from higher parts of the tree is redirected around the branch collar. The resulting irregular stem shape is limited to the area directly below the branch (Fig. 5.53).

Fluting can be caused anatomically by wide wood rays, bundled wood ray parenchyma (e.g., in beech (*Fagus sylvatica*)), or false wood rays (e.g., in hornbeam (*Carpinus betulus*)). The result is a confined transport system that locally limits the nutrient supply and radial growth and asymmetric growth rings across the entire stem area (Fig. 5.54). Studies by Schweingruber (2007) showed that cells and groups of cells can differentiate autonomously during their growth, the cambium can become locally inactive, or cork bands can interrupt diameter growth in specific areas.

Some tropical species are genetically predisposed to fluting. In extreme cases, their stem develops into a series of individual, interdependent wood segments that are perfectly matched statically. By economizing on material and increasing stability, these species gain a competitive advantage: They reduce stem volume for the sake of height growth (e.g., witte parelhout (*Aspidosperma marcgrafianum*), East Indian marking nut tree (*Semecarpus anacardium*)) (Fig. 5.55).

Flanges often have static causes. Increased radial growth at the stem butt (e.g., elm (*Ulmus laevis*), birch (*Betula pendula*), swamp cypress (*Taxodium*)) and especially in tropical species provides trees growing on soft ground greater stability. The deeply grooved and asymmetric rings affect the entire stem base. Genetics also determine the severity of the flanges.

In the tropical rain forest, nutrients are quickly washed out of the soil. Therefore, trees use their fine root system to draw nutrients from the ground directly surrounding them. They spread their roots out close to the soil surface without any deep roots to anchor them. To make up for this lack of stability, they develop buttress roots (Fig. 5.56).

Research on the anatomy of buttress roots remains limited (Comvalius 2012). It would be important from an anatomical and botanical perspective to determine how the buttress root cells are "constructed" and what gives the band of cells such a stability.

While mouldings primarily result from nutrient restrictions, flutes normally have genetic origins and flanges are linked to physical causes and genetic predisposition.

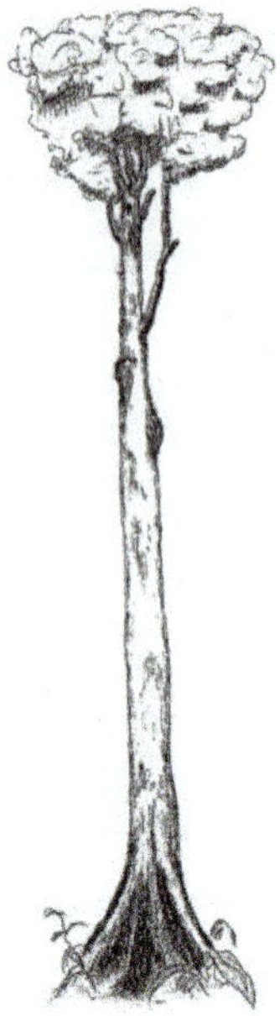

Fig. 5.56 Buttress-like flanges increase stability for a zwart riemhout (*Micropholis guyanensis*)

Fig. 5.57 Unfortunate forest scenario after poor negative selection (causing mouldings under shade branches)

5.1.4.2.3 Prevention

Selective thinning removes trees with low branching and a significant amount of *mouldings*. Forestry management practices involving stem maintenance through underwood and canopy formation promote self-pruning and reduce the number of shade branches (Fig. 5.57).

Flutes cannot be manipulated. The only way to prevent flutes in a stand is by preemptively planting species less predisposed to fluting or by selectively removing unsuitable phenotypes.

Flanges are only prevented through appropriate site selection or by removing trees prone to flanging.

5.1.4.2.4 Impact on Use

Mouldings usually result in lower wood volume recovery in most types of manufacturing. Rotary cutting produces so-called peelers (Fig. 5.58). Plane sawing results to a high amount of waste. In addition, the fiber orientation around the moulding is disrupted (Fig. 5.59).

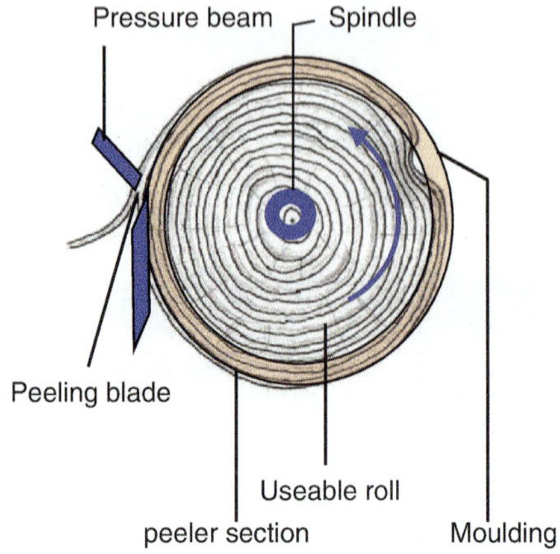

Fig. 5.58 Peelers occurring on a veneer roll from beech (*Fagus*) wood

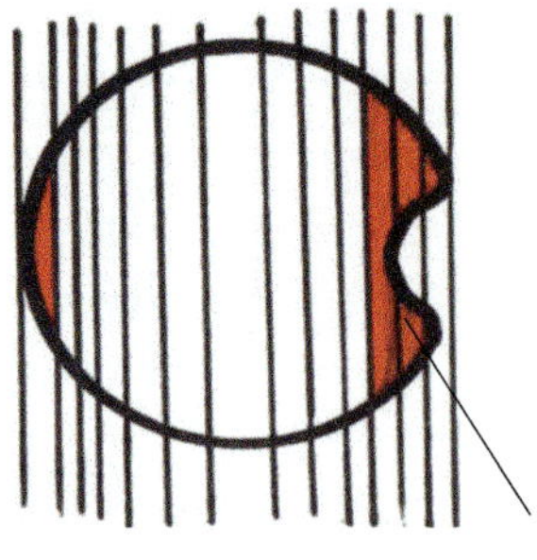

Fig. 5.59 Reduced wood recovery due to a high amount of slab wood

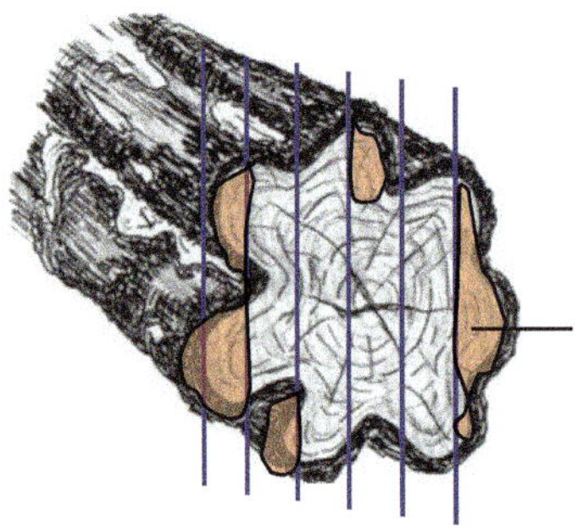

Fig. 5.60 Fluted birch (*Betula pendula*) with reduced wood recovery due to a high amount of slab wood

A moulding does not affect the otherwise superior qualities of a veneer log if the moulding is positioned at blade level.

Fluted stems are unsuitable as veneer. A large amount of slabs and splinters in plane sawing reduces the wood recovery volume (Fig. 5.60).

The asymmetric growth rings can cause the sawn timber to warp and fluted logs do not split straight.

Flanges considerably disrupt growth ring patterns at the stem base. This causes the sawn timber to warp and crack. Flanges are not suitable for veneer due to their irregular growth rings.

Although moulding, flutes, and flanges significantly reduce the applicability of logs for some uses, their impact can be largely eliminated in certain processing methods.

Nevertheless, current international timber grading standards pay little or only indirect attention to these characteristics (HKS 2002a; DIN 1997b, 1998b; RVR 2014). By contrast, in former East German grading standards (TGL 1977c), beech (*Fagus sylvatica*) veneer logs with mouldings of up to 5 cm deep were still accepted. Fluting in birch (*Betula pendula*) veneer logs was classified as fully unsuitable.

To date, only the German state of Thüringen (Thüringer Landesforstdirektion 1997) has included the impact of mouldings on the quality of beech (*Fagus sylvatica*) wood and thereby established a clear trading standard for timber sales. Based on common guidelines in the HKS, the effect of mouldings and flutes on timber can only be estimated. This leaves significant room for interpretation.

The *CEN* (DIN 1997b, 1998b does not even include the term moulding under "supplemental terms" for round wood and sawn timber terminology (DIN 2001b). This defect, however, according to common quality standards, automatically degrades a log to a quality C classification.

The German standards RVR (2014, p. 1) also generally grades all timber with moulding and fluting as average quality C class timber, unless the mouldings are compensated for by the otherwise high quality of the wood.

Possible effects of flanges on quality class are explicitly not listed in the RVR. Any quality assessments thus need to be mutually agreed upon.

5.1.4.2.5 Technological Adaptation

A *moulding* does not affect the otherwise superior qualities of a sliced veneer log if the moulding is positioned at blade level.

Crosscuts are made close to the branch collar above a moulding (Fig. 5.61). The stem should be cut using the breakdown method to ensure that the full board width is utilized (Fig. 5.62).

For the same reason, *fluted timber* should also be cut using the breakdown method. Careful stacking and drying of the sawn timber reduces the chances of warping or cracking.

Flanges need to be cut off on site or removed at the mill.

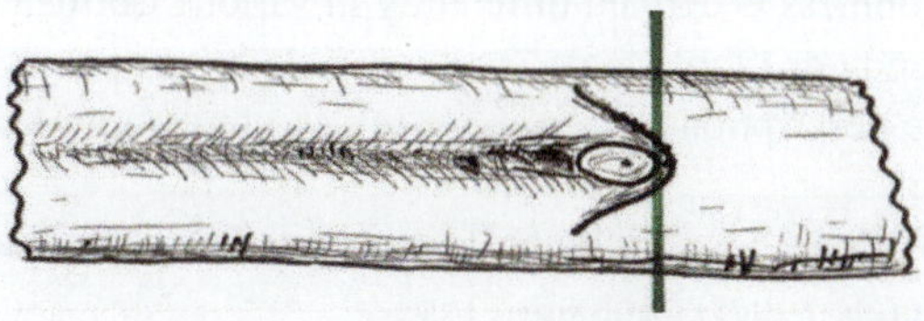

Fig. 5.61 Positioning the cut just above the branch collar with moulding

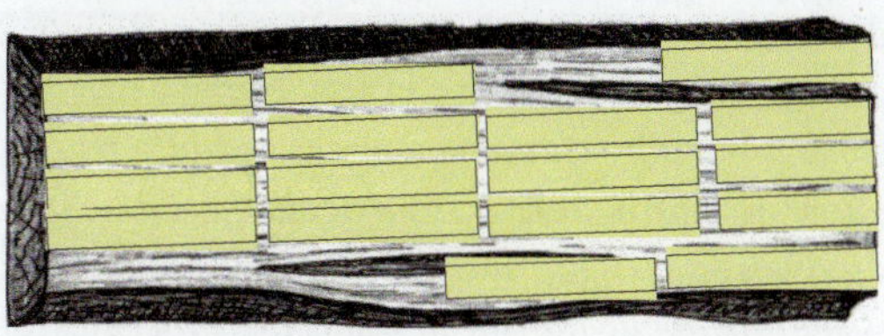

Fig. 5.62 Cutting scheme of a fluted board (breakdown cut) being cut for parquet

5.2 Limbiness

5.2.1 Live and Dead Limbs, Epicormic Shoots, and Branches

5.2.1.1 Description

Limbiness refers to all visible primary and secondary limbs on the surface of a stem, as well as all knots in the underlying wood (Fig. 5.63).

Limbs are vital parts of the tree, originating from the stem, yet distinguished from the stem wood by their own cell structure (reaction wood, pithiness, color, and density), fiber orientation, and growth rings.

Depending on where they originate, limbs are categorized as primary or secondary limbs:

Primary limbs originate as buds from the pith.

Secondary limbs (epicormic shoots/branches) are not connected to the pith. Instead, they develop from dormant or adventitious buds. The age of the secondary limb is determined based on the growth rings at the point of origin on the stem. Larger epicormic shoots (>2 cm diameter) develop into epicormic branches (Fig. 5.64).

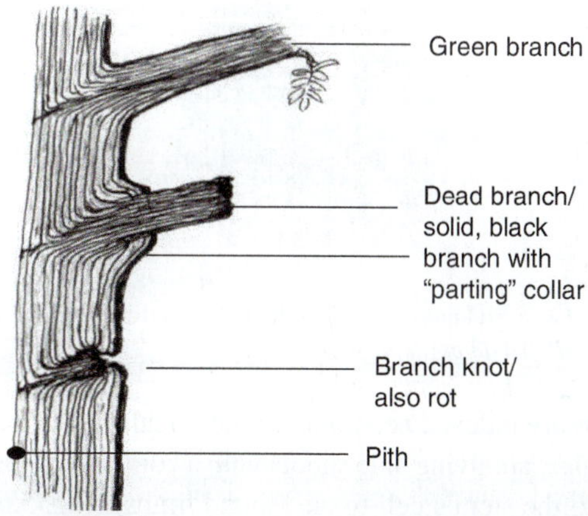

Fig. 5.63 Heartwood stem with live, dying, and knotted primary branches

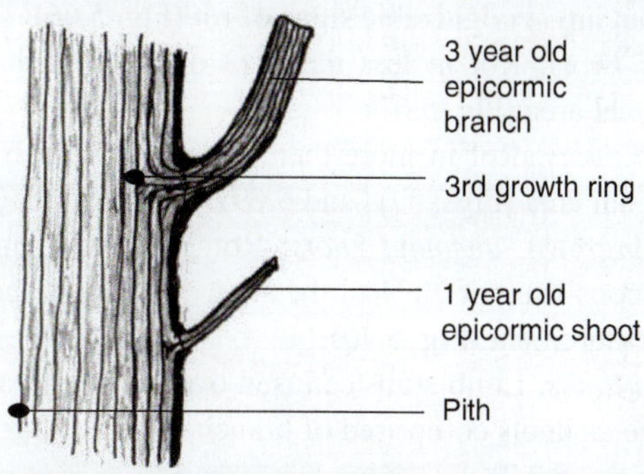

Fig. 5.64 Epicormic branch and shoot (secondary limb)

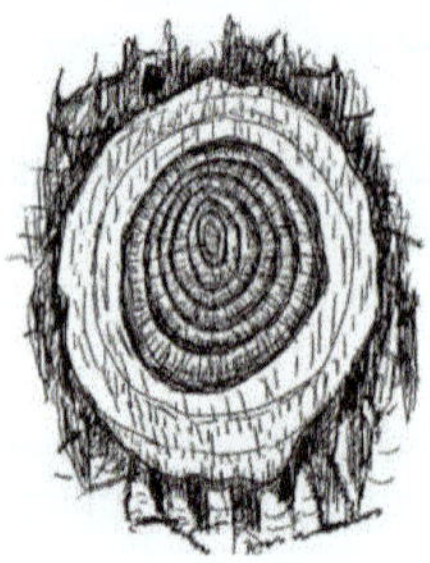

Fig. 5.65 Healthy limb in a ring porous oak tree (*Quercus robur*)

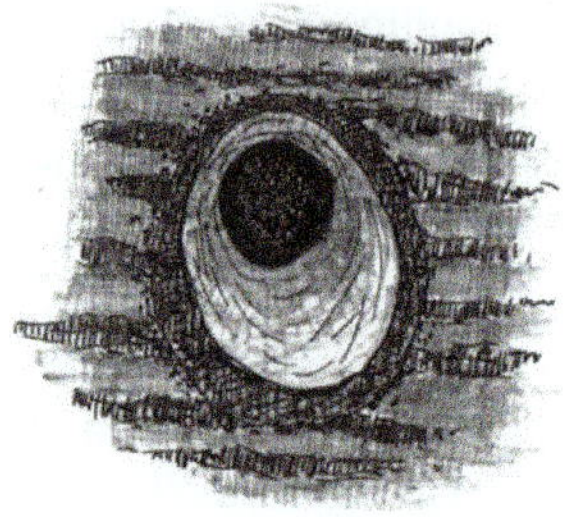

Fig. 5.66 Solid black knot in a semi-ring porous cherry tree (*Prunus avium*)

Fig. 5.67 Decayed knot: rot <1/3 in the cross-sectional area of a conifer, spruce (*Picea abies*)

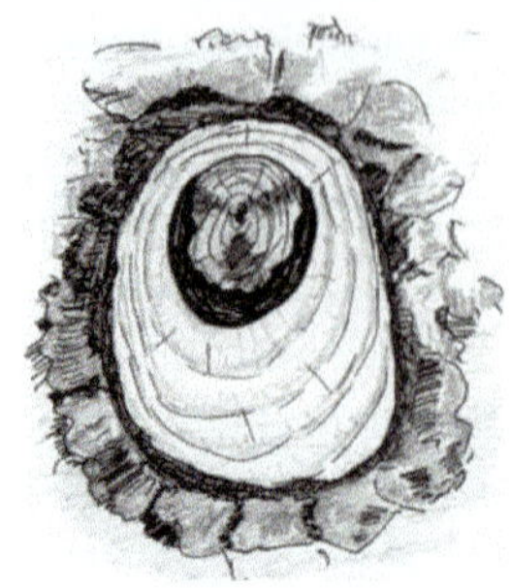

Fig. 5.68 Unsound knot: rot >1/3 in the cross-sectional area of a diffused porous tree, sycamore maple (*Acer pseudoplatanus*)

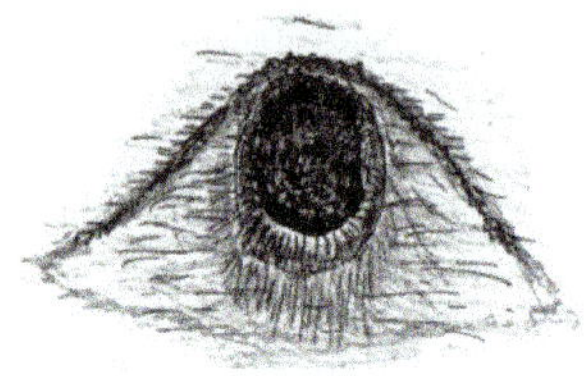

Fig. 5.69 Rotten knot in a diffused porous beech (*Fagus sylvatica*)

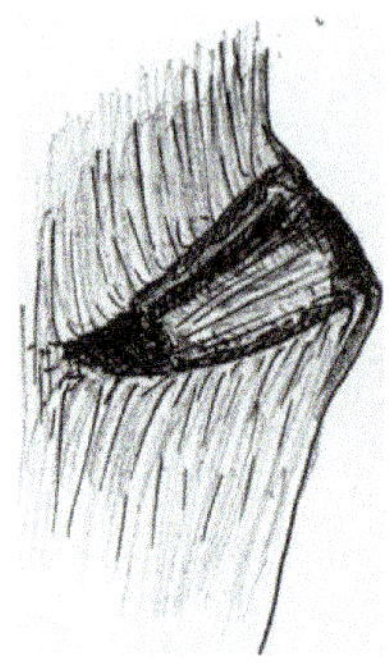

Fig. 5.70 Deeply ingrown unsound knot of a ring porous ash (*Fraxinus excelsior*)

Limbs are indexed according to their health and vitality:

Green limbs are living side shoots with a continuous connection to the stem's cell tissue (sound limbs) (Fig. 5.65) or partially isolated (dying limbs).

Dead limbs or knots are deceased limbs, no longer connected to the stem tissue.

Knots are indexed according to the degree of rot as follows:

Sound black knots: Dark-colored or black-rimmed knots without any recognizable signs of rot (Fig. 5.66)

Decayed knots: Rot in less than 1/3 of the branch cross-sectional area (Fig. 5.67)

Unsound knots: Rot in more than 1/3 of the branch cross-sectional area (Figs. 5.68 and 5.69)

Deeply ingrown, unsound knots: Rot penetration into the stem consisting of 20 % of the stem diameter at the point of measurement (Fig. 5.70)

Callused limbs: Limb stubs callused over by the stem bark, visible as knots comprised of branch seals and bark folds (see Sect. 5.2.2)

While a sound black knot consists of dead, discolored, but still sound wood (lignin turns brown, cellulose gray), the wood of an unsound knot has been fully decomposed by microorganisms and is only of limited firmness. Although most grading standards normally classify both knot types equally, from a biological perspective, they are remarkably different.

Limbiness is defined differently in various German grading standards (Table 5.5).

The descriptions given above refer to trees from temperate zones.

See Plate 5.8 for limbs as characteristic tree features (temperate zones) photos.

In principle, they apply also for tropical and subtropical species. However, given the climatic conditions in these regions they occur with much greater diversity (Richter 2012). Harzmann (1988, p. 43) notes in this context that "*...these most common quality characteristics ... usually receive little attention or no information is given about the different characteristics associated with limbiness for individual species... little is known about the species specific processes of limb decline on the stem, limb breakage and eventual callusing ....*"

See Plate 5.9 for limbiness in subtropical and tropical species photos.

Table 5.5 Limbiness characterized for de-limbed timber in various German timber grading standard

Type of limb	Definition in various grading standards			
	TGL 15799 in the former German Democratic Republic (GDR) (TGL 1977b)	HKS Grading standards for the German Federal Republic, valid until 2012 (HKS 2002a)	CEN standards of the European Union (EU) (DIN 1997b, 1998b)	German framework agreement for sawn timber trade (RVR), valid since 2014 (RVR 2014)
Healthy knot	Open knot without signs of decay	Open knot without signs of decay with less than 50 % of a black ring (SSLEF 1997). These include the heavily limbed softwoods (Hornäste) (Thuringian Landesforstdirektion 1997)	Knot with no sign of decay	"*Healthy knot: knot without any signs of decay*" Also: *"overgrown knot: Branch, which is fused with the surrounding wood on the visible side with its full cross section"*
Twigs	Started from dormant buds, thin limbs not extending to the pith	Started from dormant buds, thin limbs not extending to the pith	Twig or part of twig, visible on the stem surface	*"Nails are dead twigs with collars of less than or equal to 1 cm"*
Epicormic branches	Epicormic branches undefined	Twigs can develop into epicormic branches (Thuringian Landesforst-direktion 1997)	Twigs and epicormic branches not specifically defined	Twigs and epicormic branches undefined
Solid black knot	Dark discolored or black-edged knot without externally visible wood decay	Dark discolored or more than 50 % rimmed in black, open, and dead limb with no external sign of wood decay (Thuringian Landesforstdirektion 1997; SSLEF 1997)	Black knot, partially or fully blackened (DIN 2001b) Black-rimmed knot with more than ¾ of its area covered by bark (DIN 2001b) (equivalent to the "dead" limbs in poplar in the grading rules of the DIN 1997b?)	*Not overgrown knot: knot, which is surrounded by the visible side of a black ring or; not completely fused with the surrounding wood "usually involves a dead branch"*
Decayed knot	Open decayed knot, affecting no more than 1/3 of the cross-sectional area of the branch	Undefined	Undefined (corresponds to the "diseased and rotten limb in beech in the grading rules of DIN 1997b?)	"*Knot with visible soft rot"*
Rotten knot	Open rotten knot, affecting more than 1/3 of the cross-sectional area of the branch	Knot with externally visible rotten wood (Thuringian state forestry Directorate 1997; SSLEF 1997)[a]	Decayed knot	
Deeply rotten limb	Open rotten knot penetrating the stem by more than 20 % of the stem diameter	Undefined	Undefined	Undefined
Callused knot (callused knot based on)	Knot that is not visible on the log surface. Only apparent as a branch scar, as, for example, a blind conk, seal, or Chinese's mustach	Knot that is not visible on the log surface. Only apparent as a branch scar, as, for example, a blind conk, seal, or Chinese's mustach (TGL 1977b; Thuringian Lan-desforstdirektion 1997; SSLEF 1997)	Knot that is not visible on the log surface	*"Knots, which are not immediately visible on the surface of logs"*

[a]Frommhold (2001) defines decayed limb as "... *in addition to limbs with limited soft rot also all other limbs, including dried limbs, black knot and horn limbs with less than 50 % of its area grown into the surrounding wood"* although *in manufacturing a distinction is made between these limb types*

Beech (*Fagus sylvatica*) with branches in three life stages

Limby spruce (*Picea abies*) (dead branch retainer)

Finely limbed spruce (*Picea abies*) growing in a densely planted stand

Dying branch <1/3 of the branch diameter

Dying branch >1/3 of the branch diameter

Dead branch (solid, black branch)

Unsound knot

Assessment of an oak (*Quercus* ssp.) branch at different stages of dying

Plate 5.8 Limbs as characteristic tree features (temperate zones)

Spruce (*Picea*) with a green branch

Spruce (*Picea*) stem severed at a branch whorl with irregular growth rings and fiber orientation (Photo: G.-E. Weber)

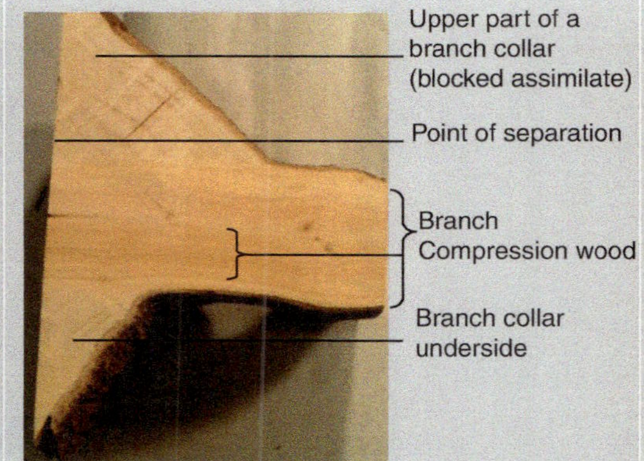

Pronounced branch collar on a living branch of a so-called "teat spruce" (*Picea abies*)

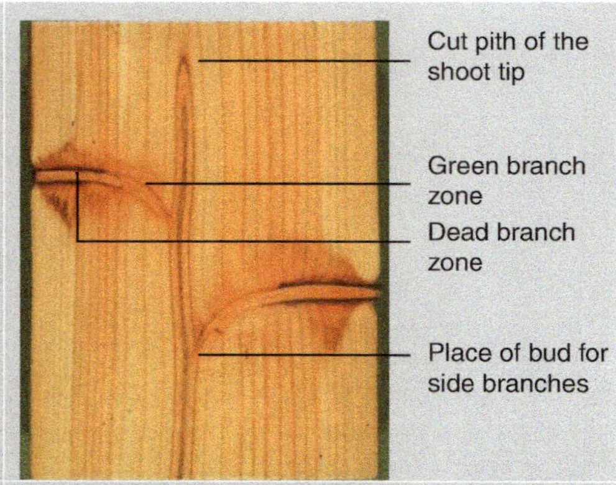

Cut pith from a spruce (*Picea*) crown with two diverging branches (inside green branch, outside dead branch)

Plate 5.8 (continued)

Extremely limby Aleppo pine (*Pinus halepensis*) in old, widely spaced stand (Troodos Mountains, Cyprus)

First population of a lava flow with Monterey pine (*Pinus radiata*). *Left*, finely limbed, *right*, heavy limbed example (Tenerife, Spain)

Water sprouts on a Monterey pine (*Pinus radiata*) (California, USA)

Water sprouts on a Norfolk Island pine (*Araucaria heterophylla*) (Australia)

Young Monterey pines (*Pinus radiata*) are pruned to prevent forest fires (Datca, Turkey)

Branch and water sprouts on a costal redwood (*Sequoia sempervirens*) after crown breakage (California, USA)

Brittle branch (brittle heart), Ø = 1. 6 m, giant redwood (*Sequoiadendron giganteum*) (California, USA)

Olive trees (*Olea europaea*), pruned for centuries and watered before harvest (Delphi, Greece)

Self-pruned eucalyptus (*Eucalyptus* sp.) (Madeira, Portugal)

Eucalyptus stem (*Eucalyptus* sp.) grows water sprouts after a fire (Madeira, Portugal)

Plate 5.9 Limbiness in subtropical and tropical species

Free-standing tree in the tropics, heavily limbed and covered with epiphytes (Surinam)

In primary forests of the tropics, only dominant trees bear a wide spreading crown (Surinam)

Rapid natural self-pruning on a Andiroba stem (*Carapa guianensis*) (Surinam)

Older group of trees in a forest, whose stems are virtually free of knots under a dense canopy (Surinam)

Exposed lamincouaru (*Minquartia* ssp.) from the middle story of a primary forest only has a small brush like crown (Surinam)

Rare: a side branch of a bolletrie (*Manilkara bidentata*) has managed to grow straight up into the canopy (Surinam)

Suddenly exposed to light, this berg gronfolo (*Qualea rosea*) quickly develops water sprouts (Surinam)

This tropical almond tree (*Terminalia* ssp.) grew branch whorls to better utilize the vertical sunlight (Surinam)

"Tree of travellers" (*Ravenala madagascariensis*) palm-similar *Strelitzie* plant with N-S-orientation of the frondleaves. The broadsides of the leaves are shined by the sun in the morning and in the afternoon (New Amsterdam, Surinam)

Plate 5.9 (continued)

Dead, but still unsealed, branch knot from the crown of a walaba tree (*Eperua* spo.) (Surinam)

Freshly sealed greenheart (*Tabebuia serratifolia*) knot continues to grow as a side branch (*center*). Large knots more commonly turn into decayed knots (*right*)

Kopi (*Goupia glabra*)

Wanakwari (*Vochysia tomentosa*)

Babun (*Virola surinamensis*)

Tropical species quickly seal off branch knots. The potential for infection, however, is so high that the area often turns into a decayed knot (Surinam)

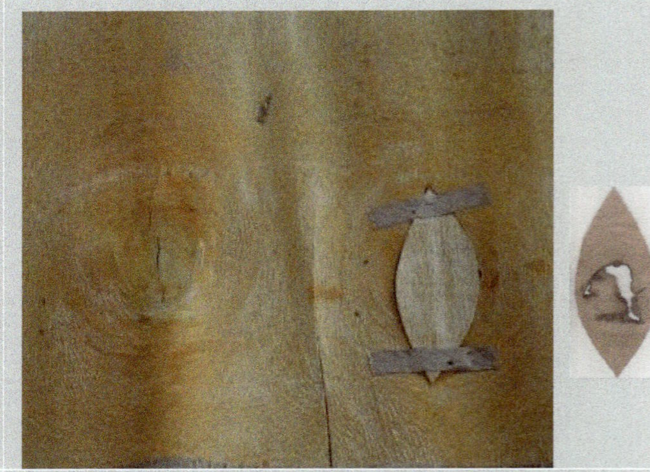

Knotty Babun veneer (*Virola surinamensis*): a black knot (*right*) is cut out and plugged. This piece of veneer is suitable as a middle layer (Surinam)

Plate 5.9 (continued)

Branch Measurement: Only the shortest distance is measured in the pithy area or the area marked by red in millimeters or the border area clearly marked by discoloring and growth rings, not, however, the breaking point, sap wood, and branch collar (Fig. 5.71) (TGL 1977a; RVR 2014; SSLEF 1997).

Assessing Stem Limbiness: Limbs have the greatest impact on the practical value of wood from temperate zone species. On average, 68 % of beech (*Fagus sylvatica*), 87 % of spruce (*Picea abies*), and 63 % of fir (*Abies alba*) logs are degraded due to limbiness (Grammel 1989). Therefore, great emphasis is placed on assessing limbiness during scaling. All grading systems use the same method for measuring limb size but assess the existence and condition of limbs differently. Although a certain degree of defect was accepted in the TGL 15799 of East Germany – a limb was tolerated if the stem was in otherwise good condition (TGL 1977a) – this is no longer acceptable in the HKS (2002b) and CEN (DIN 1997b, 1998b) unless agreed to by the customer. The change in defect tolerance reflects a shift away from accepting wood as product of nature towards considering it a technically standardized object.

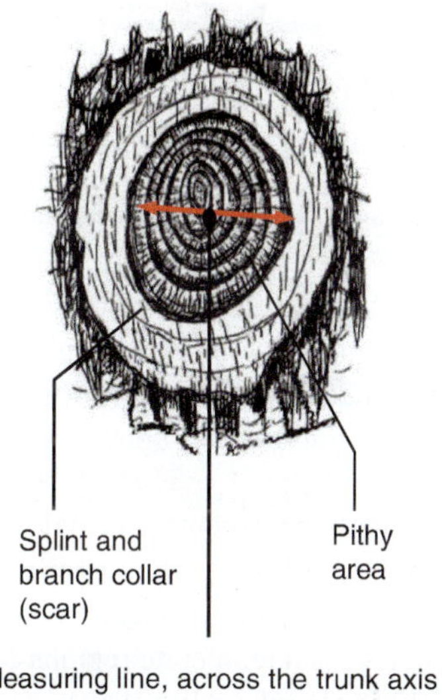

Fig. 5.71 Pruned branch showing points of measurement

However, according to RVR in the "*general rules for grading logs*" (RVR 2014, S. 17/20) "*…some defects, which do not meet the appropriate class standards,* [including limbs] *can be balanced out by the otherwise high quality of the wood….*"

The move from measuring only the number of limbs per linear meter as in HKS (SSLEF 1997) to including limb diameter sums per linear as in CEN (DIN 1999a) led to stricter grading practices. A study of grade assessments for the four main tree species showed that 22 % of all classifications based on the more stringent grading rules for limbiness, as established in the CEN, resulted in a degradation of the wood (Kotte 2002).

Although the RVR (2014) foregoes the impractical diameter measurements in favor of the simpler HKS practice, it has a stricter margin for defect that tends to lead to further degrading; as shown in this classification of a fictitious oak stem (Fig. 5.72):

Wood sorted in this example according to CEN standards is degraded sooner than wood sorted based on HKS rules. This applies particularly for the accepted nails and roses in the grade class A, the accepted number of limbs in class B, and the abrupt degrading of decayed limbs from classes C to

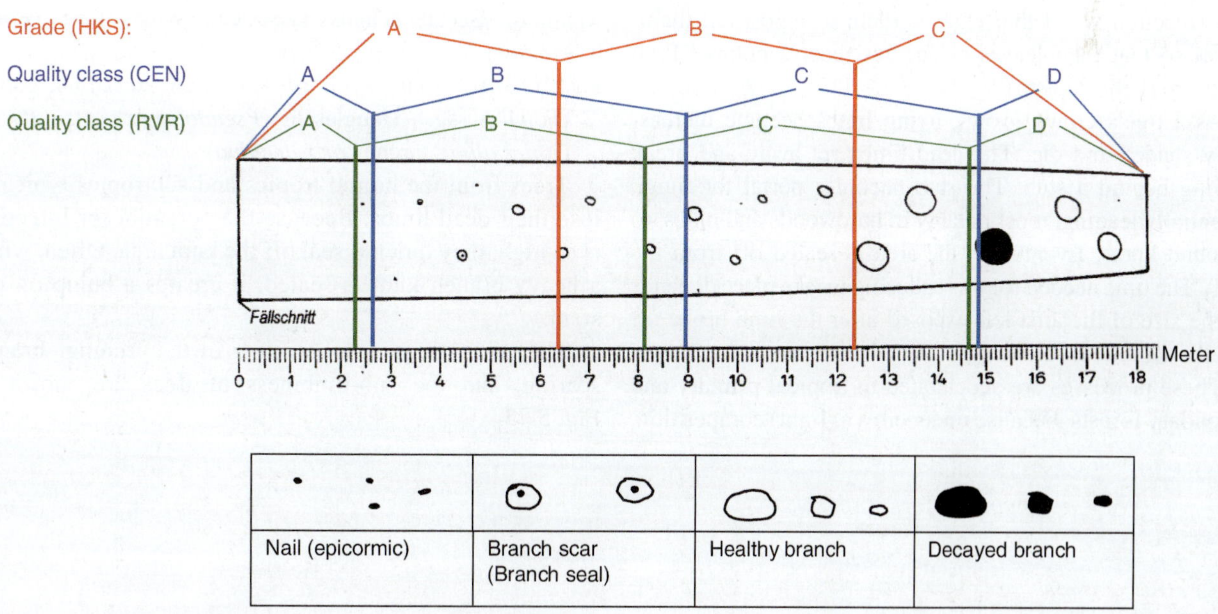

Fig. 5.72 Proposal for grading an oak stem based on limbiness according to HKS (*red*), CEN (*blue*), and RVR (*green*)

Table 5.6 Change in grading quality of logs from the state forest in Baden-Württemberg between 1953 and 1994 (Mahler 1995)

		Proportion [%]	
Tree species	Grade	1953 (*Picea abies* 1955)	1994
Spruce, fir, Douglas fir (*Picea abies, Abies alba, Pseudotsuga menziesii*)	A	0	0
	B	99	77
	C/D	1	23
Pine (*Pinus sylvestris*)	A	10	2
	B	70	15
	C/D	20	83
Beech (*Fagus sylvatica*)	A	12	1
	B	76	42
	C	12	57
Oak (*Quercus* ssp.)	A	10	1
	B	73	26
	C/D	17	73

D. The new German RVR (2014) eliminates the cumbersome branch sum measurement rules of the CEN but maintains the requirements on the quality classes A and B. This has been an ongoing trend since the 1950s and reflects the different points of view held by market stakeholders over the long term (Table 5.6).

From a technological point of view, the trend is reversed: In this case, wood sold as C and D quality can be processed using modern technology for higher-quality purposes, such as for veneer plywood liners, scantling glued wood, and finger joints.

5.2.1.2 Causes

Limbs support the assimilations organs, leaves, and needles. Thus, they are vital elements of the tree. Starting as buds, they shoot out at varying angles from the tree stem. They form reaction wood that enables them to reposition themselves so that their leaves or needles receive optimal light exposure (heliotropism).

As a tree's crown grows, living limbs become increasingly shaded and die. The dead limbs rot away and break leaving behind a stub. The stub becomes portal for fungi, potentially leading, most notably in hardwoods and pines, to unsound knots. Eventually, the area is sealed off from the stem. The time needed for this callusing to take place depends on the size of the area left exposed after the limb broke off and the vitality of the tree.

These processes are accelerated in tropical primary and secondary forests, because under crown (light) competition, the shaded limbs quickly die. Dead limbs fall off and the knot is quickly sealed to prevent infection from spreading.

See Plate 5.9 for limbiness in subtropical and tropical species photos.

Hardwoods – with the exception of wild cherry and poplar – tend to shed dead limbs, while conifers usually retain them, even when they are fine limbed.

Knots in species from temperate regions callus in the following stages (Knigge and Schulz 1966):

Slow: Birch (*Betula pendula*), hornbeam (*Carpinus betulus*), maple (Acer ssp.), pine (*Pinus sylvestris*), spruce (*Picea abies*)

Medium: Beech (Fagus sylvatica), white pine (*Pinus strobus*)

Fast: Oak (*Quercus* ssp.), ash (*Fraxinus excelsior*), poplar (*Populus* ssp.), Douglas fir (*Pseudotsuga menziesii*), fir (*Abies alba*), larch (*Larix decidua*)

Trees from the humid tropics and subtropics typically lose their dead limbs. Because the potential for infection is so high, they quickly seal off the knot area. Often, when a heavy branch knot is sealed, it creates a bump on the stem.

The main stages of branch growth, pruning, branch scaring, and the sub-limbiness of trees are shown in Fig. 5.73.

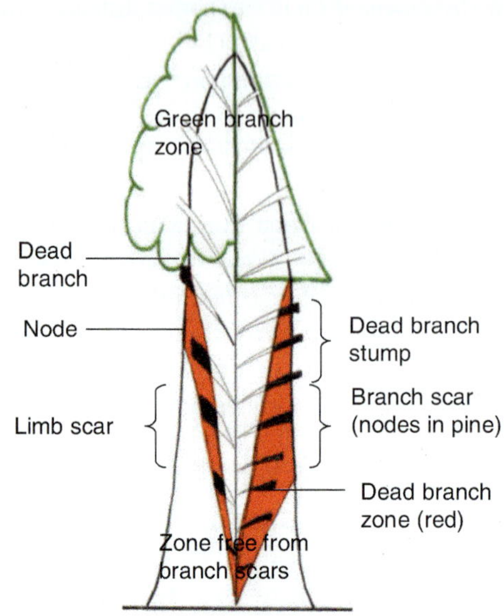

Fig. 5.73 Basic branch development, self-pruning, and limbiness among hardwoods (*left*) and softwoods (*right*)

Fig. 5.74 Large oak (*Quercus* ssp.) stem shielding a beech (*Fagus sylvatica*) sapling

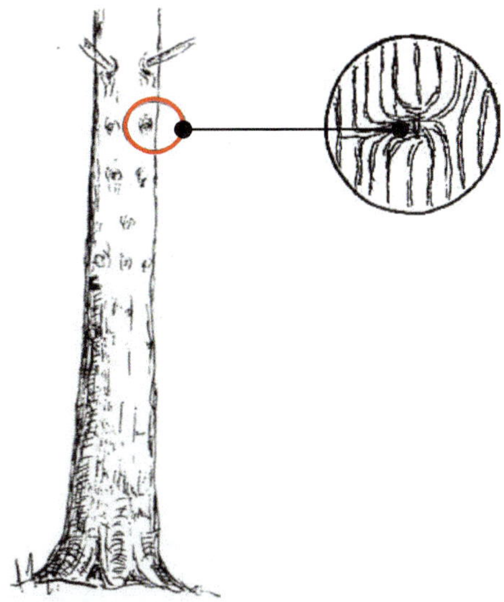

Fig. 5.75 Pruned spruce *(Picea abies)* heals quickly

5.2.1.3 Prevention

Limbiness *cannot be prevented, only minimized*. For the tree, having many limbs means having a large area of assimilation. How these limbs are distributed along the stem and whether they are thick or thin depend on the growth dynamics of the tree.

In commercial forests, silvicultural practices can influence branch distribution and thickness:

- A *high-density stand at the planting and seedling stages* results in fine limbs in the brush stage and early self-pruning.
- Silvicultural practices for stem maintenance such as understory planting and species diversity promote self-pruned, high-quality stems and deter epicormic shoots (Fig. 5.74).
- Finely branched phenotypes are preferable for stand formation and should be promoted during stand maintenance.
- Selected conifers should be pruned green (Douglas fir (*Pseudotsuga*), larch (*Larix*), and spruce (*Picea*) or dead spruce (*Picea*)) during the vegetation period. Among hardwoods, only poplars and aspens (*Populus tremula*) require pruning. To ensure healthy scaring, limbs should be cut in front of the branch collar, preferably stumped, and then cut again later (Fig. 5.75).

Pruning should occur when the time for natural self-pruning has passed (at the latest when one third of the target diameter is reached). Pruning is particularly recommended when wide spacing between trees leads to heavy limbiness which, left unpruned, would detract from the quality of the wood. This is often the case in tree plantations.

Research on Douglas firs (*Pseudotsuga*) studying the dependency between pruning heights and pruning methods and recoverable timber value and investment risk have shown that manual pruning with pneumatic cutters up to a height of 6 m brings the best economic returns (Küthe 1986).

High density stands in the youth, linked to fine limbiness, are generally preferred over widely spaced, less expensive stands, because finely branched timber generates higher sales revenue.

See Plate 5.10 for stem maintenance and pruning photos.

300-year-old oak (*Quercus petraea*) stem protected by a beech understory (Spessart, Germany)

Pruned spruce (*Picea abies*) (*right*) next to a Scotch pine (*Pinus sylvestris*) (*left*) (Selb, Germany)

Douglasie (*Pseudotsuga menziesii*) after the second pruning (Vogelsberg, Deutschland)

Plantation of *Acacia auriculiformis* before the first pruning (Quang Tri Province, Vietnam) (Photo: Le Thien Duc)

Plantation of young Teak wood (*Tectona grandis*) after the first pruning (Ghana) (Photo: C-T-Bues)

Log from pollarded spruce (*Picea abies*) (Fichtel Mountains, Germany)

Fully sealed knot in an ash (*Fraxinus excelsior*)

Plate 5.10 Stem maintenance and pruning

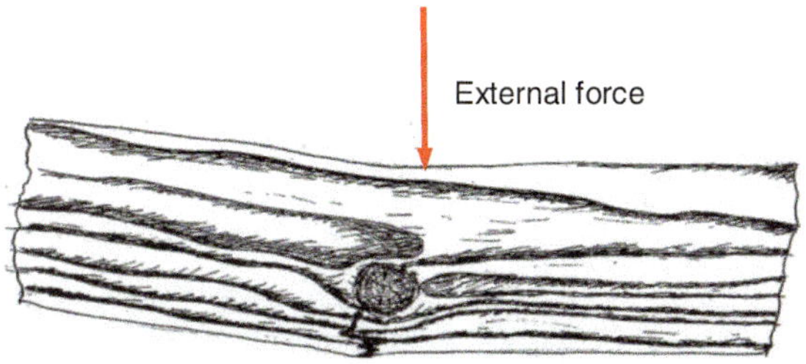

Fig. 5.76 Break in a square edged log on the tension side of a dry branch after bending pressure

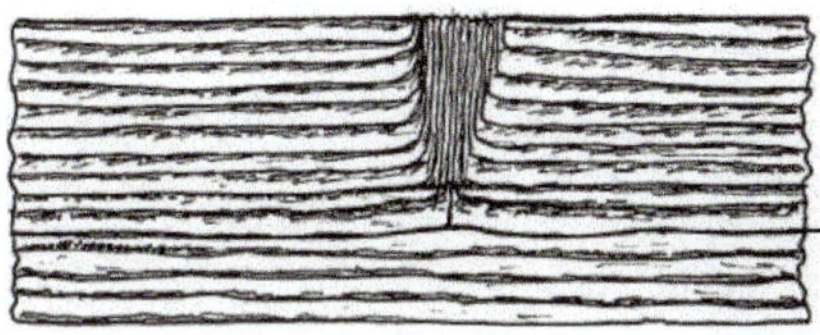

Fig. 5.77 Wing branch on a radial cut board surface

5.2.1.4 Impact on Use

A limb is a combination of features. It consists of the following basic characteristics that may affect the use of its wood (Richter and Bues 2003a):

- Variations in the chemical composition of wood (high lignin content, "black-score"-coefficient) reduce the pulp yield and quality.
- Variations in the anatomical structure (growth ring construction, reaction wood) reduce the strength of timber products. Therefore, construction timber must be oversized (DIN 2003), (Fig. 5.76).
- Variations in the fiber orientation complicate surface processing (rough wood surface near the branch, Fig. 5.77).
- Surface roughness causes variations in color and finishes absorption.
- Isolation from surrounding wood can turn a dead branch into a loose knot thereby reducing the practical value of the sawn timber (Fig. 5.78). As construction wood, a branch easily reaches breaking point.
- Limbs have varied swelling and shrinkage properties based on the higher density. This leads to cracks during drying (Fig. 5.79).
- Ingrown bark, especially bark folds in the collar, reduces timber yield and quality.

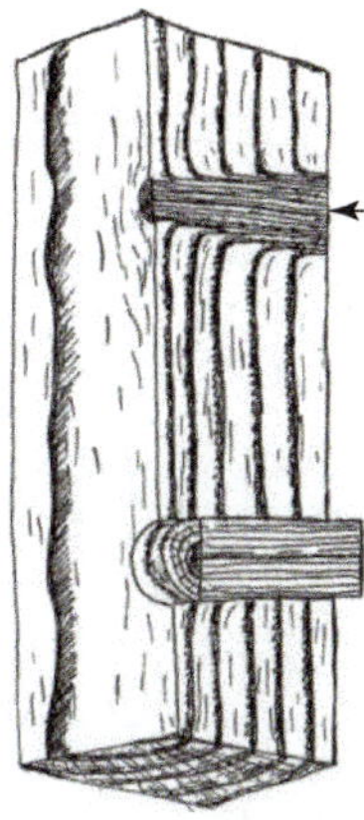

Fig. 5.78 Anatomical differences between live and dead (lower) branches

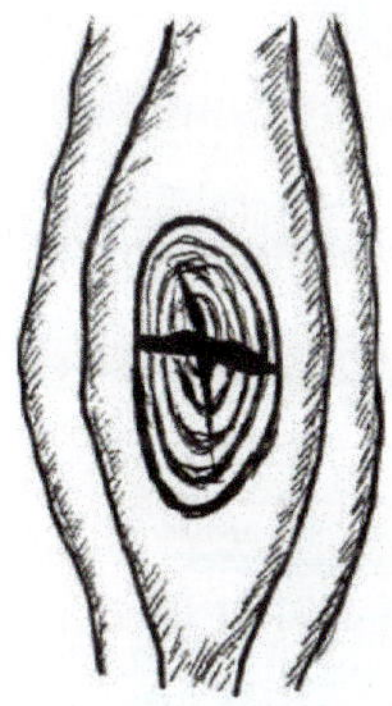

Fig. 5.79 Dry crack in a knot in a board

- Resin accumulation (compression wood zone in conifers) effects durability of paints.
- Fungal infection further reduces branch stability and discolors the branch wood.
- Discoloring (reaction wood, amount of latewood) can detract or add to the wood's appearance. Branches represent natural and unique wood characteristics often used to create distinctive wood pieces such as the knotty wood furniture made from common pine (*Pinus sylvestris*) or Swiss pine (*Pinus cembra*).

See Plate 5.11 for photos of the branch as desired and undesired characteristic and Plate 5.12 for photos of the branch as a design element and constructive feature.

Pine (*Pinus sylvestris*) boards with sound knot (links) and dead branch (*right*)

Pine (*Pinus sylvestris*) board with broken black knot

Spruce (*Picea abies*) board with loose knot

Black knots ("nails") on larch board (*Larix decidua*)

Rough fiber caused by planning against the grain, spruce (*Picea abies*)

Enclosed bark in narrow-angled branch, pine (*Pinus sylvestris*)

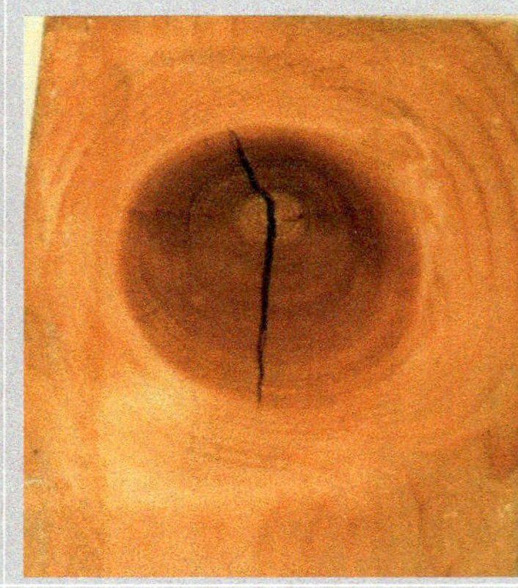

Sound knot with dry crack, spruce (*Picea abies*)

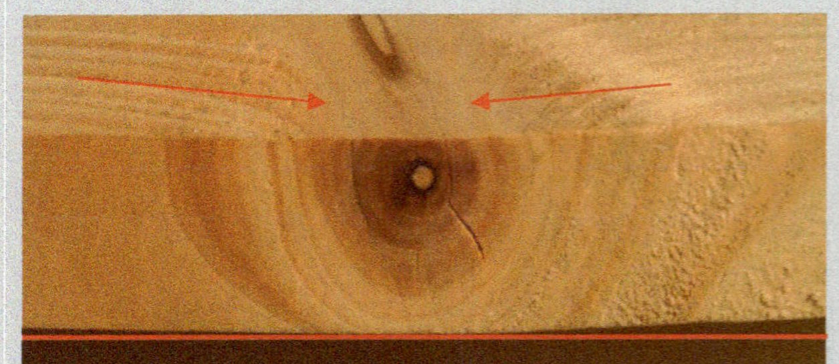

Crooked spruce (*Picea abies*) board due to heavily shrunk splay knot

Plate 5.11 The branch as desired and undesired characteristic

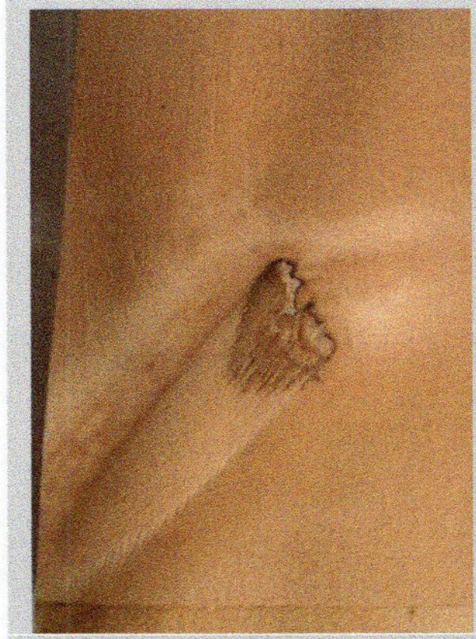

Maple (*Acer* ssp.) veneer with slash knot

Elm (*Ulmus glabra*) veneer with loose knot

Plugged loose knots in a composite spruce (*Picea abies*) panel

Elm (*Ulmus glabra*) veneer with epicormic shoots scars

Yew (*Taxus baccata*) veneers with epicormic shoot scars

Plate 5.11 (continued)

Old log house made with spruce (*Picea*) with interesting branch features (Simplon, Switzerland)

Rustic furniture made of limby juniper wood (*Juniperus*) (Ketchum, Idaho, USA)

Old door from knotty Swiss pine (*Pinus cembra*) boards, Graubünden, Switzerland

Kitchen cabinet made of knotty pine (*Pinus sylvestris*)

Frames and bows from branch collars for the replica of a Viking ship (Roskilde, Denmark)

Replica of a Viking ship. The mast retainer is made from crooked branches (Roskilde, Denmark)

Harp-shaped spruce (*Picea abies*) used in Viking ship replica, Gauldalen, Norway

Plate 5.12 The branch as a design element and constructive feature

Ziegelheimer crucifix. Body and head from the trunk and branch of a lime tree (*Tilia* ssp.) mid-fifteenth century (Waldenburg, Germany)

Standing desk made from a knotty chestnut (*Castanea sativa*) (Tharandt, Germany)

Trough and spicket made from extremely limby mountain spruce (*Picea abies*) (Allgäu, Germany)

A woodworker's selection of troughs for gardening and landscaping (Allgäu, Germany)

Plate 5.12 (continued)

Table 5.7 Impact of limbiness in the RVR standards

Species		RVR quality classes			
		A	B	C	D[a]
	Limb type[b]	Healthy knots	Healthy knots, as well as black and rotten limbs	Healthy knots, as well as black and rotten limbs	Healthy knots, as well as black and rotten limbs
Spruce/fir[c]	Healthy, overgrown	Unacceptable	≤4 cm	≤8 cm	Acceptable
	Not overgrown	Unacceptable	≤3 cm	≤6 cm	
	Rot	Unacceptable	Unacceptable	≤3 cm	
Pine[c]	Healthy, overgrown	Unacceptable	≤5 cm	≤8 cm	
	Not overgrown	Unacceptable	≤5 cm ???	???	
	Rot	Unacceptable	Unacceptable	???	
Larch[c]	Healthy, overgrown	Unacceptable	≤5 cm	≤8 cm	
	Not overgrown	Unacceptable	≤3 cm	≤7 cm	
	Rot	Unacceptable	Unacceptable	≤3 cm	
Douglas fir[c]	Healthy, overgrown	Unacceptable	≤5 cm	≤8 cm	
	Not overgrown	Unacceptable	≤4 cm	≤7 cm	
	Rot	Unacceptable	Unacceptable	≤4 cm	
Beech	Healthy	Unacceptable	2 per 4 m, ≤10 % d_M	Accepted in normal size)[d]	Acceptable
	Rotten	Unacceptable	1 per 4 m, ≤10 % d_M	2 per 4 m, ≤20 % d_M, max. 12 cm	
Oak/red oak	Limbs				
	Healthy ≤4 cm Ø	Unacceptable	2 per 4 lfm	Accepted	
	>4 cm Ø	Unacceptable	1 per 4 lfm	Accepted	
			or		
	Rotten ≤4 cm Ø	Unacceptable	1 per 4 lfm	Accepted	
	>4 cm Ø	Unacceptable	Unacceptable	2 per 4 lfm	
	Twigs (individual)	1 twig r/2 m	Accepted	Accepted	

[a]40 % of the wood volume must be scalable for grade D
[b]Epicormic branches are classified as healthy limbs
[c]Softwood standards are not yet ratified; therefore, the amended 2010 "normal" or "limited" refers to the general definition of quality class
[d]As regards the general description of the quality classes

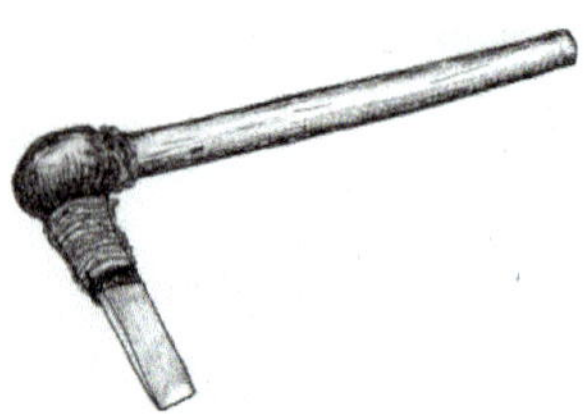

Fig. 5.80 Ötzi's ax handle (prehistoric man found in the Ötztal Alps on the boarder of Austria and Italy (Fleckinger and Steiner (2000))

Impact of limbiness on quality class based on the Framework Agreement for Timber Trade in Germany (RVR 2014) is shown in Table 5.7.

5.2.1.5 Technological Adaptation

There are various ways to minimize the negative effects of limbs on wood products or even to take advantage of the limbiness. Our prehistoric ancestors used branches as handles for their flint and bronze axes (Fig. 5.80). Wood cutters have long known the benefits of branch whorls when splitting logs (Fig. 5.81).

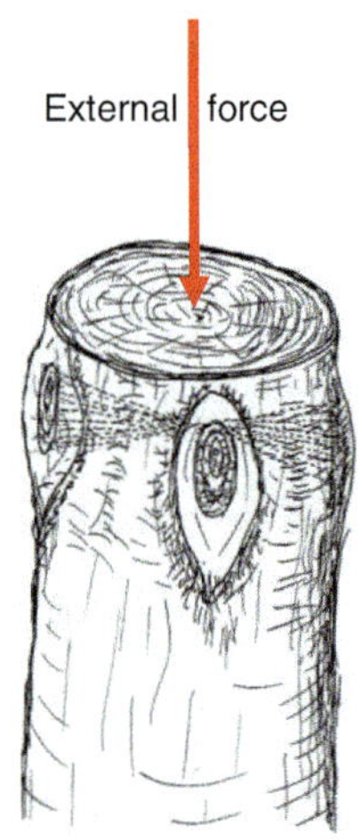

Fig. 5.81 Branch whorl helpful in splitting logs (Dowel effect)

It is advantageous to divide a veneer log (figuratively) into sections. Branches and limb scars should be positioned as closely as possible to the blade or in sections C and D in order to achieve the highest recovery from sections A and B (Fig. 5.82).

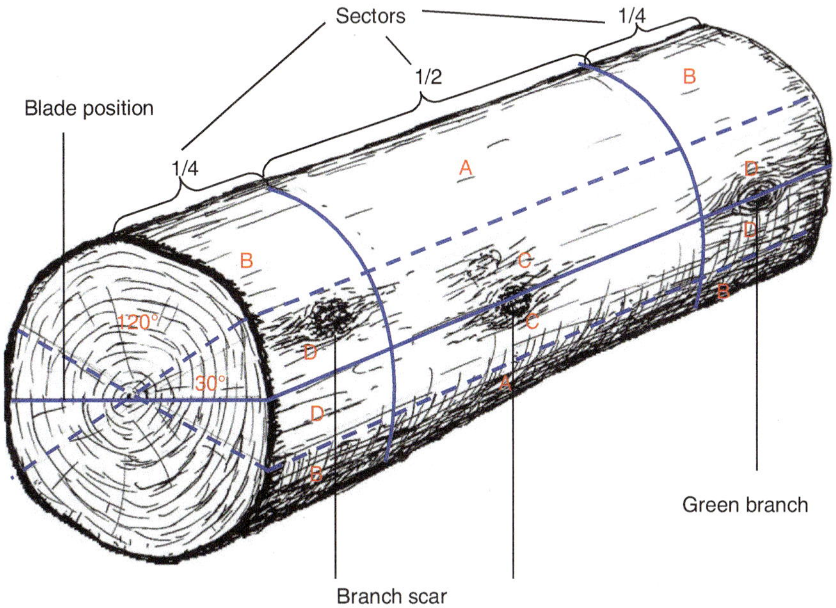

Fig. 5.82 Determining blade position and cutting scheme for a veneer log. *Legend: Sector A*: most valuable part of a veneer bolt. *Sector B*: exterior of a veneer block. *Sector C*: interior of a log. *Sector D*: exterior of a log

Epicormic shoot scars are common in some tree species such as yew (*Taxus baccata*) and poplars (*Populus* sp.). They produce highly figured veneer.

Construction wood should be mechanically tested for strength to determine the weakening effect of any limbs.

In the production of composite lumber, the varying strengths of the individual wood particles are balanced by gluing them together.

Unwanted limbs in the sawn timber can be removed with a drill and filled with a plug.

5.2.2 Limb Scars

5.2.2.1 Description

Of all the wood characteristics, limbiness causes most quality downgrades in timber assessments. As early as 1954, Taffé found that the impact of branches on quality grade and market value had an 87.5 % influence on spruce (*Picea abies*) timber (Taffé 1954), and Schulz (1961) found a 68.2 % influence for beech (*Fagus sylvatica*), cited from Grammel (1989). Knots in the underlying wood typically have significant consequences for the wood's technical properties (Rast 1982). It is therefore important through the use of specific diagnostic procedures to determine the depth and size of the knot along with the angle of the living limb to the stem. For this purpose, limb scars are an excellent diagnostic tool. This applies to trees in temperate as well as tropical zones.

A *limb scar* is a callused branch or a twig stub left behind on the stem. It is identified by the structural changes of the bark in the scarred area. Each limb scar consists of a collar and a ridge (in contrast to bark injuries).

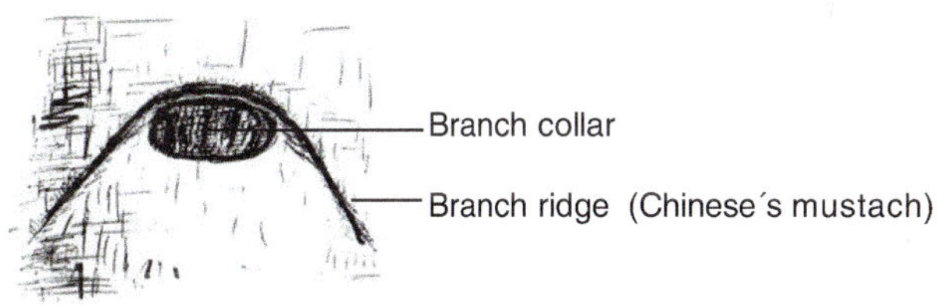

Fig. 5.83 Limb scar in a smooth-barked beech (*Fagus sylvatica*)

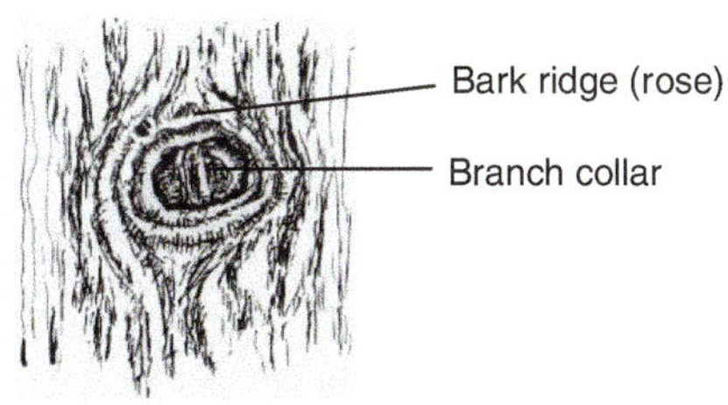

Fig. 5.84 Limb scar in a rough-barked oak (*Quercus* ssp.)

In species with smooth bark, such as beech, birch, maple, hornbeam, linden, aspen, and alder *(Fagus sylvatica, Betula pendula, Acer* ssp., *Carpinus betulus, Tilia* ssp., *Betula pendula, Alnus* ssp.*)*, the ridge runs down both sides of the limb stub resembling a beard, thus earning the nickname "Chinese's mustach" (Fig. 5.83). Ridges and collars can take on different shapes, depending on the angle of the former living limb to the stem (Schulz 1958).

On roughly textured bark species, as, for example, oak (*Quercus* ssp.), a more or less circular-shaped ridge forms around the stub often called a *rose* (Fig. 5.84).

Limb scars on rough-barked softwoods such as certain pine species (*Pinus* ssp.), larch (*Larix* ssp.), and spruce

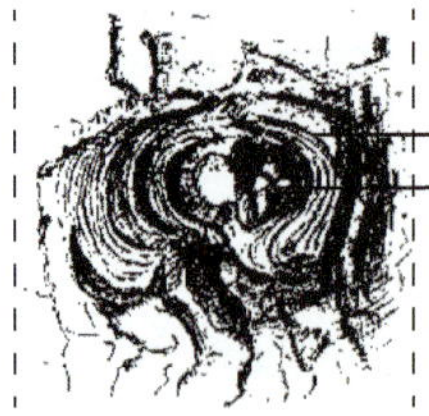

Fig. 5.85 *Larch (Larix decidua)* with limb scar

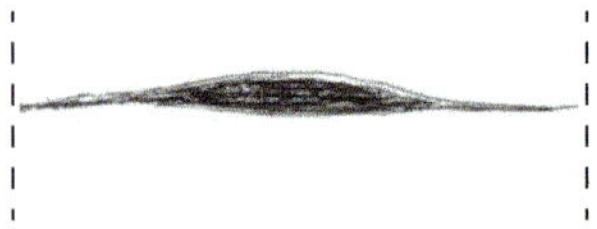

Fig. 5.86 Bark strip in an ash *(Fraxinus excelsior)*

(*Picea* ssp.) are similar to those found on rough-barked hardwoods (Fig. 5.85). Limb scars on smooth-barked softwoods such as fir (*Abies* ssp.), as well as most young softwoods, are similar to those found on smooth-barked hardwoods.

A *bark strip* is a long, narrow, hardly recognizable bark ridge. It originates from a branch formed during the tree's juvenile period, for example, in beech (*Fagus sylvatica*) and ash (*Fraxinus excelsior*) (Fig. 5.86).

A *blind conk is* a pronounced swelling of the stem over a scarred stub (Fig. 5.87).

Dead epicormic shoots leave behind small limb scars the size of nail heads in the bark and often remain attached to the stem as "twigs" (larch (*Larix* ssp.)). They are also called *nails* (Fig. 5.88).

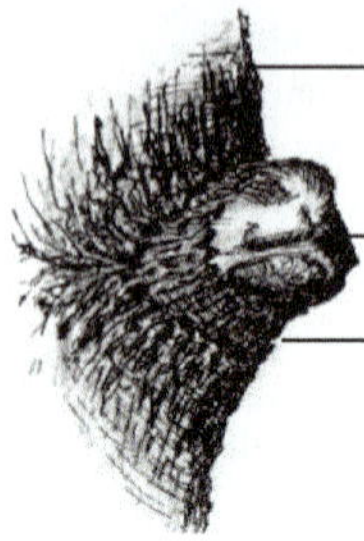

Fig. 5.87 Blind conk in an ash (*Fraxinus excelsior*)

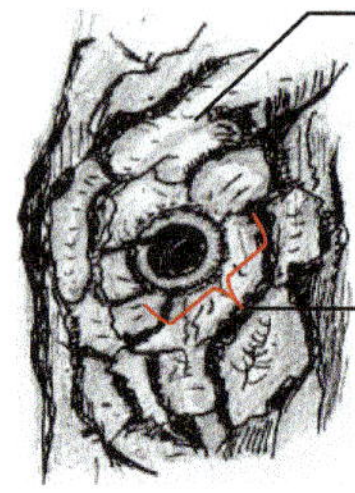

Fig. 5.88 Limb scar from twig (nail), oak (*Quercus* ssp.)

See Plate 5.13 for limb scars on softwoods, Plate 5.14 for limb scars on ring porous hardwoods, Plate 5.15 for limb scars on diffuse porous and semi-ring porous, and Plate 5.16 for limb scars on tropical hardwoods.

Spruce (*Picea abies*) with barely visible limbscars on three branch whorls

Spruce (*Picea abies*) with older branch scars (branch whorls)

Limb scars on two branch whorls in an old pine (*Pinus sylvestris*)

Limb scar on a pine (*Pinus sylvestris*) with recently sealed branch collar

Limb scar on a Colorado fir (*Abies concolor*)

Barely visible limb scar on a Weymouth pine (*Pinus strobus*)

Old, wide limb scar on a larch (*Larix decidua*)

Nail (epicormic shoot stub) on a larch (*Larix decidua*)

Two old limb scars on a Douglas fir (*Pseudotsuga menziesii*) (longitudinal cracks in the branch collar)

Limb scar on a Douglas fir (*Pseudotsuga menziesii*) about 5 years after pruning

Limb scar on a fir (*Abies homolepis*) with inverted bark ridges (hanging branches!)

Limb scar on a false cypress (*Chamaecyparis lawsoniana*)

Plate 5.13 Limb scars on softwoods

Limb scar on an oak (*Quercus* ssp.)

Older "roses" on an oak (*Quercus* ssp.). The branch collar already with longitudinal cracks

Younger "roses" on an oak (*Quercus* ssp.). The branch collar sealed off the branch stub

Asymmetric limb scar on an oak (*Quercus* ssp.). The branch received left-sided sunlight exposure

Old "rose" from an epicormic shoot on an oak (*Quercus* ssp.). The branch collar is cracked longitudinally

Fresh "nail" on an oak (*Quercus* ssp.). The epicormic shoot lived for 3 years

Older limb scar on a red oak (*Quercus rubra*). Above and below bark strips

Robinia (*Robinia pseudo acacia*) with old limb scar

Left: ash (*Fraxinus excelsior*) with old limb scars. *Center*: limb scars on a younger ash (*Fraxinus excelsior*) with bark strips. Not to be confused with ash bud moth (*Hylesinus* ssp, *Leperisinus* sp.)scars!

Right: young ash (*Fraxinus excelsior*) with new limb scars and unsealed knots (*left*)

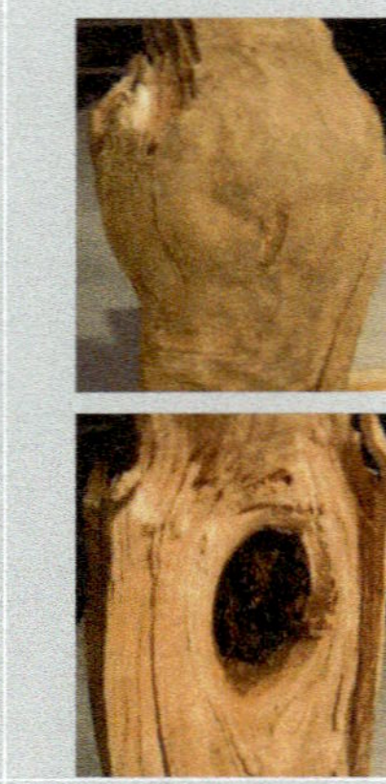

Branch knot on an ash (*Fraxinus excelsior*). Frontal view healthy (*above*). Hidden inside is a massive dead knot (*below*)

Plate 5.14 Limb scars on ring porous hardwoods

The height of the Chinese´s mustaches on a beech (*Fagus sylvatica*) indicate the varying angles of the former branches

"Scar face" on overgrown stubs of extremely slanted branches on a beech (*Fagus sylvatica*)

Bark strips in the lower stem section–wide limb scars from twigs of the once young beech (*Fagus sylvatica*)

Bulging branch knots on a lower storey beech (*Fagus sylvatica*) with limited growth in diameter

Common maple (*Acer pseudo-platanus*) with two limb scars

Old limb scar on a Norway maple (*Acer platanoides*)

Old limb scar on a birch (*Betula pendula*)

Narrowly angled branches leave behind long Chinese´s mustaches on this birch (*Betula pendula*)

Limb scars on a nearly 30-year-old aspen (*Populus tremula*)

Limb scars on a ca. 10-year-old aspen (*Populus tremula*)

Bark strips on an aspen (*Populus tremula*)

Bulging branch knots on a dead poplar (*Populus* sp.)

Plate 5.15 Limb scars on diffuse porous and semi-ring porous hardwoods

Older limb scar (*below*) and open decayed knot (*above*) in a linden (*Tilia cordata*)

Recently sealed branch stub in a hornbeam (*Carpinus betulus*)

Alder (*Alnus glutinosa*) with bark ridges shaped like triangles

Almost sealed branch collar on a young alder(*Alnus glutinosa*)

Limb scar on a walnut (*Juglans regia*) with a "nail" on the top *left* and to the *right* an epicormic shoot (semi-ring porous)

Limb scar with freshly sealed branch collar in a wild cherry (*Prunus avium*) (semi-ring porous)

Plate 5.15 (continued)

Massive sumaruba stem (*Simarouba amara*) in a tropical primary–forest knot-free in the lower bole section (Surinam)

Barely visible limb scars in a bolletrie (*Manilkara bidentata*) in a tropical primary forest (Surinam)

New limb scar on a free-standing coral tree (*Erythrina glauca*) in a coffee plantation (Surinam)

Free standing alley tree with twigs over a limb scar (Surinam)

Newly sealed knot on a zwart riemhout (*Micropholis guyanensis*) (Surinam)

In primary forests trees are normally topped at harvest below the first decayed knot (kopi (*Goupia glabra*)) (Surinam)

The 3 m knot free stem section above the branch stub in a gronfolo (*Qualea rosea*) log is a bonus at no extra cost for the sawmill (Surinam)

Above: walaba (*Eperua* ssp.) with decayed knot
Below: when cut, the knot in the underlying wood is visible

Locust copal (*Hymenaea courbaril*) with well-sealed branch stub

Due to its diameter of 2 m intended for export. Extremely knotty (*Terminalia dichotoma*) (Surinam)

Plate 5.16 Limb scars on tropical hardwoods

Yellow Lapacho log (*Tabebuia serratifolia*) with knots before and after cutting (*right*: cut open knot) (Surinam)

Gronfolo (*Qualea rosea*) with a large decayed knot and moulding below

Scarred branch stump on this kopi (Goupia glabra) significantly reduce timber yield

Massive wanakwari (*Vochysia tomentosa*) knot (cut from a log)

Plate 5.16 (continued)

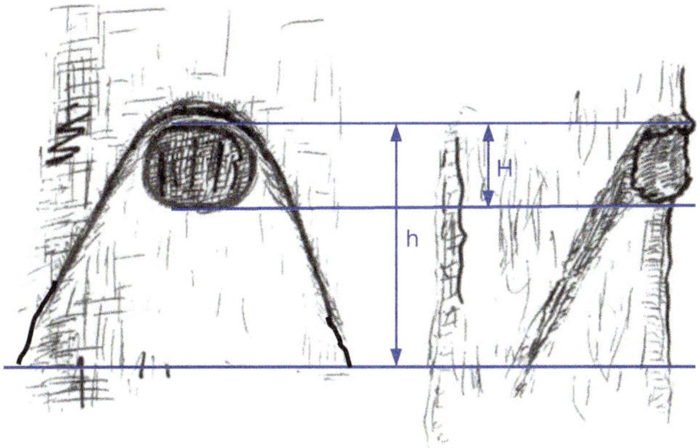

Fig. 5.89 Measuring the heights of a bark ridge (Chinese's mustaches) *h* and limb scar *H*

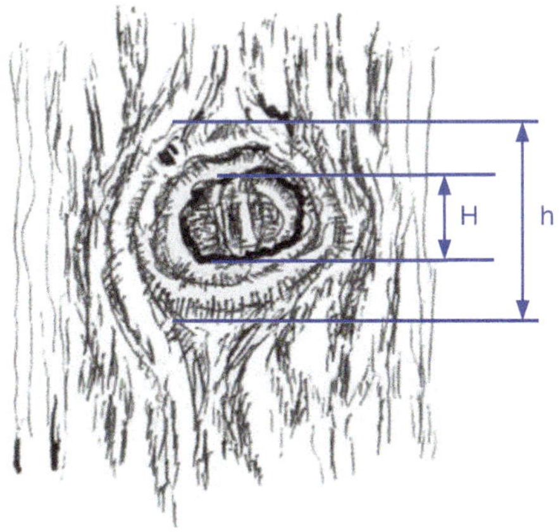

Fig. 5.90 Measuring the exterior diameter of a bark ridge (rose) *h* and height of the branch collar *H*

Measurement (Frommhold 2001; Mette 1965; SSLEF 1997):

- *Bark ridges in smooth bark species (Chinese's mustaches):* The measuring range (h) is the distance between the line connecting the ends of the Chinese's mustach and the top of the limb scar (Fig. 5.89).
- *Bark ridges in rough bark species (rose):* The distance (h) is the outermost edge of the limb scar, measured vertically down the stem (Fig. 5.90).
- *Branch collar* (*H*): The boundaries of the branch collar, measured vertically down the stem (Figs. 5.89 and 5.90).
- *Nail*: In Mette (1965), nails of up to 2 cm diameter are considered epicormic shoot or twig collars; all others are referred to as branch collars. Currently, the standard accepted collar height is up to 1 cm (Frommhold 2001).

Limb scars are measured only in the direction of the stem's axis since they tend to expand horizontally as the stem grows in diameter. In the European standards (CEN), neither the "procedures for measuring dimensions" (DIN 1998a) nor the "measurement of characteristics" (DIN 1998b) includes the measurement of limb scars.

Assessing limb scars on stems: A limb scar is simply a bark feature indicating the presence of a knot in the underlying wood. Therefore, its impact on processing the wood is rather insignificant.

Branch collars and branch ridges, however, particularly in smooth-barked, so-called "honest" tree species, provide a good indication of how deeply the knot is embedded in the wood, the strength of the living limb, and the angle of the limb to the stem. This diagnosis is also possible in rough-barked species, but subject to greater uncertainty.

Probably the first quantified studies between limb scars and limb geometry were conducted by Wakin, Polubojarinov, and Solovyov (Wakin et al. 1969) and included in the 1972 Soviet GOST timber grading standards (GOST 2140–71). These guidelines permitted the depth of knots in birch (Betula) to be determined based on the angle of the Chinese's mustach.

Rast (1982) quantified the depth of the knots based on the shape of the branch collar in red oak (*Quercus rubra*). Nevertheless, the view continues to persist in current forestry practice and various grading standards that the most effective way to determine the amount of clear wood in a stem is to measure the ratio of height to width of Chinese's mustaches (Erteld and Achterberg 1954; König 1957, Wagenführ and Scheiber 1989). This is wrong! (Richter and Bues 2003b).

The depth of the knot in the underlying wood is measured as follows: The height of the branch collar (*H*) has the same relationship to the diameter of the stem at the time when the limb died off (*2r*) as the width of the branch collar (*B*) has to the current stem diameter (*2R*). This proportion is based on the supposition that the limb (collar) height and the limb (collar) width are equal at the time when the limb died off.

On the basis of radiation set, it is possible to determine the radius of the stem at the time the limb broke off (*r*) to stem with the current radius (*R*) through the relationship of height (*H*) to width (*B*) of the branch collar (Bosshard 1984b; Knigge and Schulz 1966) (Fig. 5.91 frontal view and top view):

$$r \sim \frac{H \times R}{B}$$

The *size of the limb* at the knot *BSt* is about half of the branch collar *H* (Erteld and Achterberg 1954):

$$BSt \sim \frac{H}{2}$$

In the Soviet GOST standard (GOST 2140–71 (1972)), this average was generally for stem diameters of 24–28 cm (close to the limb) at a branch collar height of 10 cm. Subsequently, the limb size increased disproportionately with an increase in stem diameter and branch collar height. That means, for example, that a 40 cm thick stem with a branch collar height of 16 cm would correlate to a probable branch strength of 12 cm, equal to 75 % of the branch collar height.

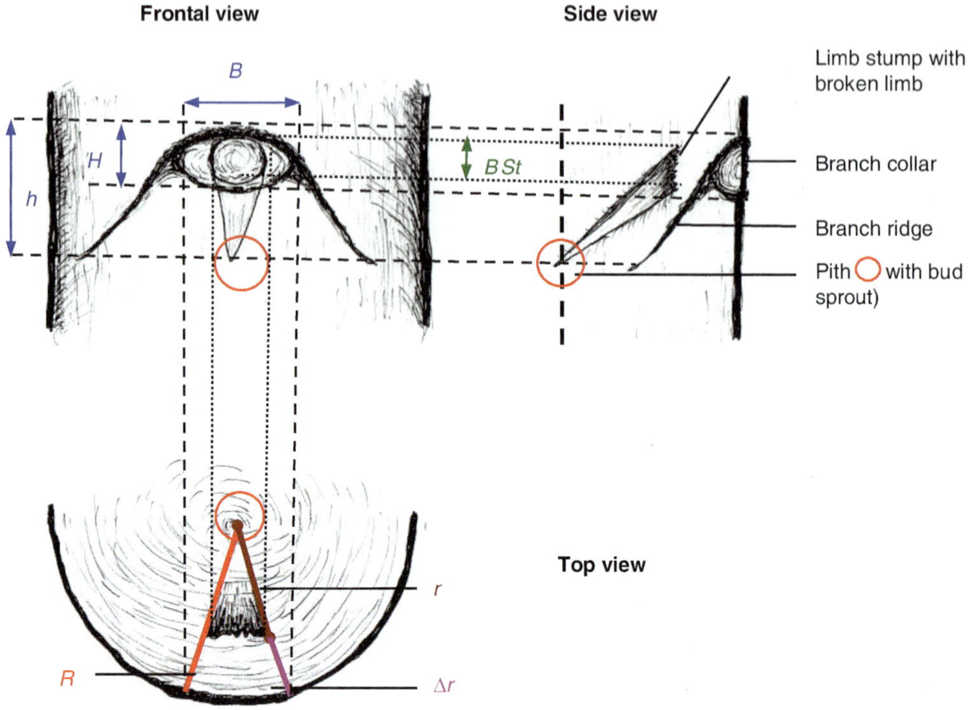

Fig. 5.91 Different views of a scarred branch stump in a beech stem (*Fagus sylvatica*)

Table 5.8 Varying rates of scarring among species

Scarring speed	Hardwoods	Softwoods
Slow	Birch, hornbeam, maple (*Betula pendula, Carpinus betulus, Acer* ssp.)	Pine, spruce (*Pinus sylvestris, Picea abies*)
Medium	Beech (*Fagus sylvatica*)	White pine (*Pinus strobus*)
Fast	Oak, ash, poplar (*Quercus* ssp., *Fraxinus excelsior, Populus* ssp.)	Douglas fir, fir, larch (*Pseudotsuga menziesii, Abies alba, Larix decidua*)

The *angle of the limb* is determined from the height of the bark ridge *h*. The measuring point lies on the line between the ends of the Chinese's mustaches and the top of the branch collar. The stem cylinder is penetrated by the ingrown knot (Fig. 5.91, side view).

5.2.2.2 Causes

As a limb grows, it forms a branch ridge. This ridge continues to increase in size as long as the branch continues to grow in diameter. If the branch dies and breaks off, it is encircled by a "parting collar." This collar eventually grows over the stump of the dead branch.

This overgrowth is called a branch collar.

The rate of scarring differs among species (Table 5.8) (Knigge and Schulz 1966):

In *smooth-barked trees*, the living limb pushes the branch ridge (Chinese's mustach) higher as it grows out of a horizontal position. This is common among birch (*Betula pendula*).

The bark ridge expands as the stem diameter increases but maintains the same height as it had when the dead limb broke off (Fig. 5.92).

In *rough-barked trees*, the bark ridge encircles the swollen branch stub (Fig. 5.90). The bark turns itself up into a rose. The rose completely surrounds the branch collar.

In tropical primary and secondary forests, most limbs do not live long. Crown competition, irradiation angle, and angle of sunlight combined with insufficient light exposure rapidly kill off shade branches (see Sect. 3.3). The area left behind is quickly sealed without any obvious scarring. On the other hand, heavy branches that break after withstanding long periods of crown competition usually leave behind an area too large for the tree to immediately seal off resulting in a high risk of infection and often leading to decayed knots that leave behind significant protrusions on the tree stem (Fig. 5.93).

5.2.2.3 Prevention

Limb scars cannot be prevented, only limited in number and size. Dense spacing is recommended during the planting and seedling stages and should be maintained until the clear stem bole achieves a desired length. In commercial forests, management practices, such as understory planting and tree species diversity, promote natural self-pruning resulting in fewer limb scars.

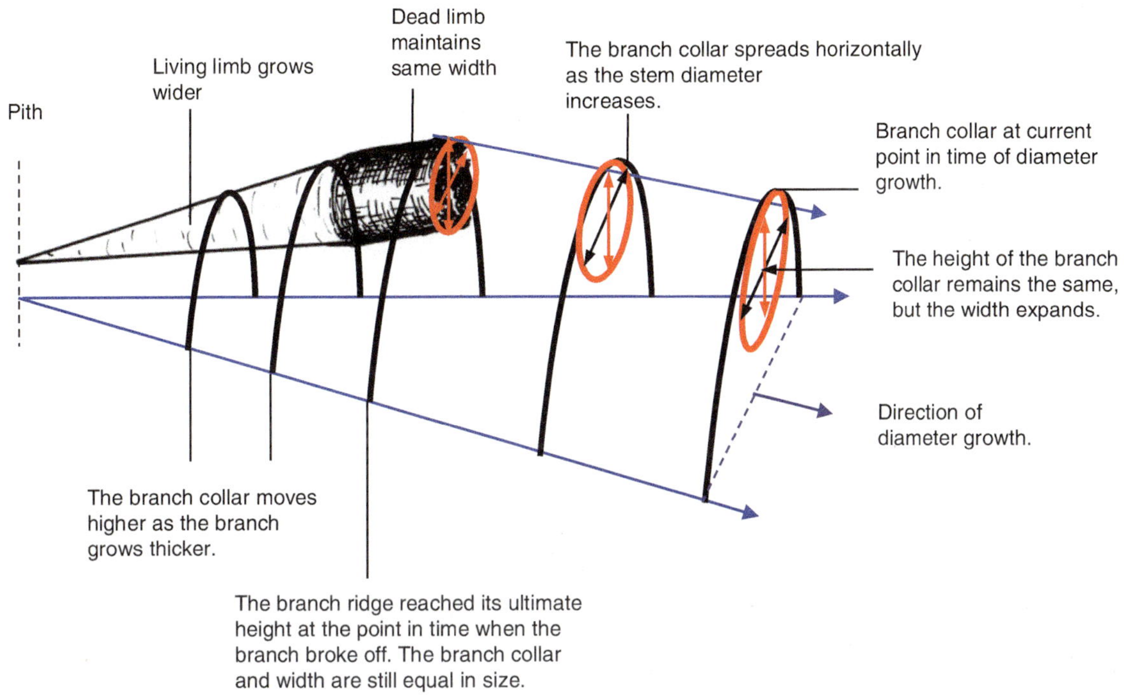

Fig. 5.92 Schematic diagram of a growing, dying, broken off, and scarred branch and the resulting changes in the bark ridge and branch collar expansion

Fig. 5.93 Walaba (*Eperua* ssp.) branch bulge with decayed knot

Finely limbed phenotypes should be promoted. When possible, selected trees should be pruned green or dry so that the limb will die off quickly and with less scarring.

Trees in tropical primary and secondary forests experience heavy shading in the lower stem sections resulting in a process of "natural pruning" – a main reason for the above-average quality of the wood from these regions with regard to the knottiness. In plantations, however, additional pruning measures beyond the natural processes are often necessary due mainly to the geometric spacing of the planted trees.

5.2.2.4 Impact on Use

Limb scars are important external diagnostic features. The limb scars themselves do not affect wood processing. Depending on the intended use of the barked timber, scars can be viewed as either decorative features or wood defects. Knots are the most problematic.

The impact of limb scars is assessed based on the current RVR (2014) standards as follows (Table 5.9):

5.2.2.5 Technological Adaptation

In wood processing, it is often important to know the thickness of the limbless wood (Δr). With peeled veneer, this information helps determine the amount of knot wood in the scrap roll. With sliced veneer, knowing the amount of knot wood allows for a more effective cut thereby ensuring maximum yield. The same applies to the processing of sawn timber. The following is an example for calculating the amount of limbless wood (Figs. 5.94 and 5.95). The equation is

$$\Delta r = R - \frac{H \times R}{B}$$

Table 5.9 Impact of limb scars on quality class based on RVR standards (RVR 2014)

		RVR quality class			
Species scar type		A	B	C	D
Spruce/fir (*Picea/Abies*) Pine (*Pinus*) Larch (*Larix*) Douglas fir (*Pseudotsuga*)		The RVR gives no information on the admissibility of limb scars and their tolerable dimensions. Strictly speaking, that means that limb scars/roses are accepted even for A quality wood if all parties agree, as according to the *general grading rules, that they do not detract from the otherwise good quality of the wood*			
Beech (*Fagus*)	Limb scar	Accepted if the scar (branch collar to width) is ≤1:4	Accepted if the scar (branch collar to width) is ≤1:2 and scar height is ≤ 10 cm	Accepted	Accepted
Oak (*Quercus*)	Limb scar (rose, nail)	1 rose per 1 m ≤3 cm branch collar height	Roses: 4 per 4 m ≤4 cm branch collar height nail. unlimited	Unlimited	Unlimited

Fig. 5.94 The branch collar on this piece of beech (*Fagus*) stem is double as wide (*B*) as it is high (*H*)

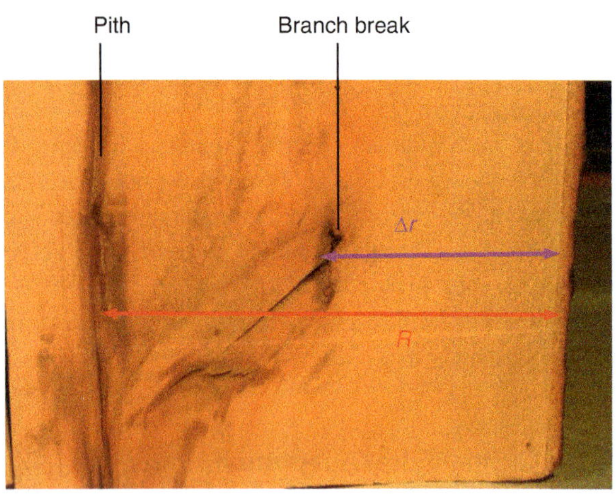

Fig. 5.95 The center cut of this stem piece (from Fig. 5.94) shows that its branchless wood (Δr) is almost half as wide as its radius (R)

Fig. 5.96 Irregular tree ring structure in a cross section of a 170-year-old ring porous oak sessile oak (*Quercus petraea*) (Germany)

Fig. 5.97 Erratic tree rings both within and between rings in a spruce cross section (*Picea abies*) from a reforested field in Germany

5.3 Anatomical Structure

5.3.1 Irregular Tree Rings/Increment Zones

5.3.1.1 Description

A tree ring is the ring of new wood formed in the cambium each year during the vegetation period. In climates with seasons, trees experience alternating periods of growth (summer) and dormancy (winter). As a result, tree rings form with varied degrees of color and wood density. In temperate climates, variations in tree ring patterns characterize the tree's growth process over the course of its lifespan (Fig. 5.96).

Irregular tree ring structures are represented by significant width variations within individual rings or between the tree ring widths of the stem cross section (Fig. 5.97). Site conditions (climate, soil) affect the tree ring widths. At their northern limit, for example, pines, junipers, and birches can have tree ring widths of less than 0.1 mm. Trees in subtropical and tropical climates, on the other hand, can form increment zones up to 2 cm wide. These increment zones differ from tree rings in that they form dependent on periods of rainfall and drought. Trees in tropical evergreen rain forests exhibit only indistinct increment zones because their growth is continuous. Only trees from temperate climates which form annual tree rings will be discussed further.

Tree ring width variations influence wood density differently in softwoods than in hardwoods. Softwoods with wide tree rings have a relatively low wood density due to their wide-lumined earlywood tracheids, while softwoods with narrow tree rings have a higher density because of their lignin-rich latewood tracheids. In diffuse porous hardwoods, on the other hand, tree ring width variations have little effect on wood density. Ring porous hardwoods form only a few wide-lumined cell rows, and as the tree rings grow wider, so does the portion of narrow-lumined latewood. This leads to higher density (Mette 1984; Schweingruber and Isler 1991; Wagenführ and Scheiber 1989).

A tree's age can be determined by studying its *annual tree rings*. The influence of dry years and extreme events, temperature history, amount of sunlight exposure, and nutrient supply are recorded in the annual ring widths (Hartig 1856; Knuchel 1934, 1954).

The average annual ring width is measured by dividing the *outer* 75 % of the radius (R) of the weaker end of a stem section (butt end) by the number of rings (n) (RVR 2014). This method eliminates the influence of irregularities in tree ring structure at the stem base and the above-average widths of the inner rings in juvenile wood.

$$\phi \text{ tree ring width} = \frac{R\ (75\%)[mm]}{n\ (75\%)}$$

Dendrochronology can determine the exact year a tree ring was formed. This is done by synchronizing or cross-dating the chronological sequences of tree ring features from a species within a larger growing region. The data can be used, for example, to precisely date wood objects (Delorme 1973; Douglass 1919; Heussner 1994; Huber 1941; Polge 1963; Schweingruber and Isler 1991; Schweingruber 1995; Schweingruber 2012).

Dendro-archeo-morphology, a multivariate cross-correlation involving tree ring analysis, dendrochronology, and tree morphological studies, and including internal and external tree characteristics, can be used to derive complex conclusions on a tree's growth history as well as on how the wood was handled and processed before becoming a finished object (Richter 2008a, b).

See Plate 5.17 for tree ring as a diagnostic tool.

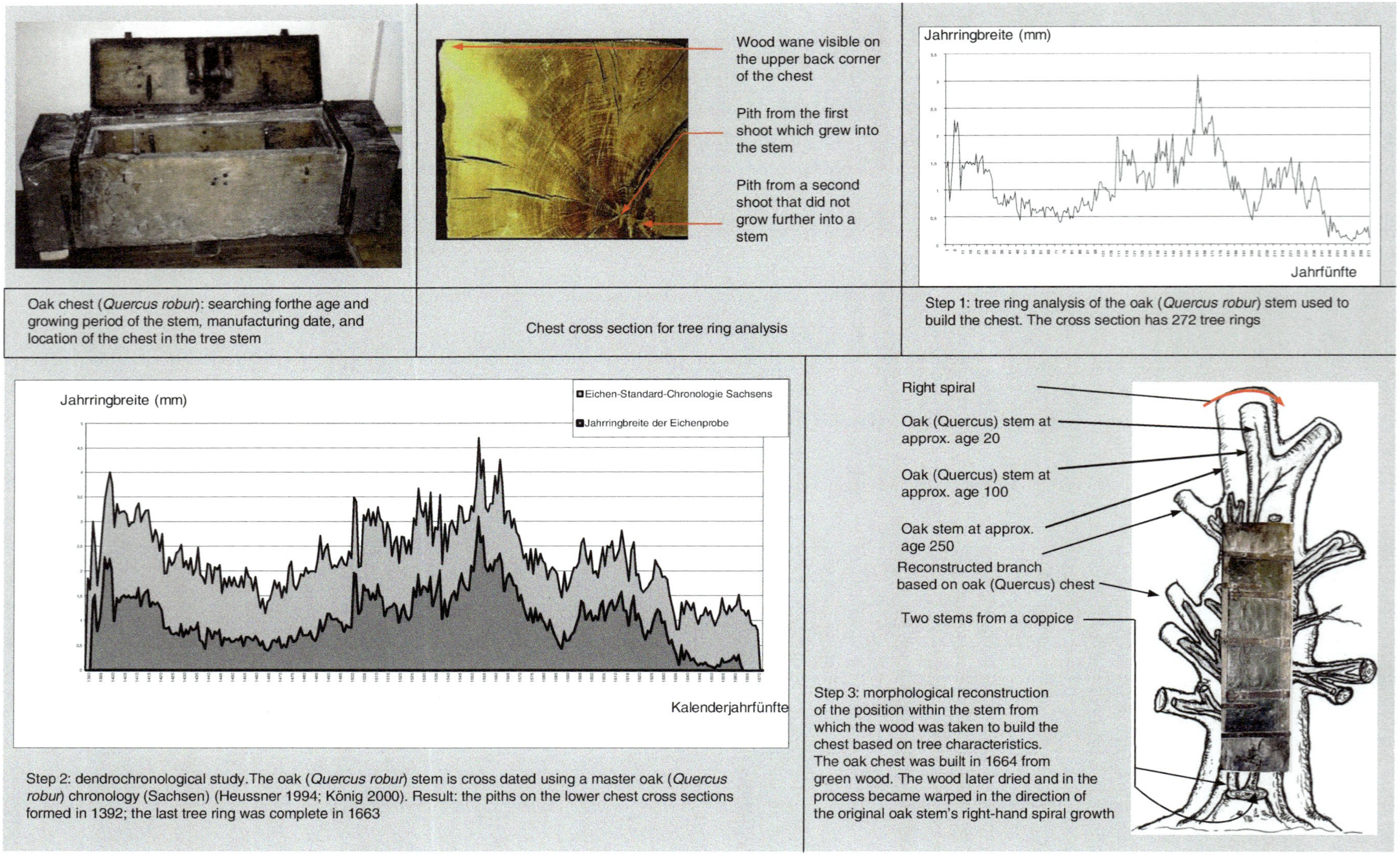

Plate 5.17 Tree ring as a diagnostic tool: dendro-archeo-morphological studies on an ancient oak chest (*Quercus robur*) (Richter et al. 2004)

Fig. 5.98 On the cross section of algarrobo (*Prosopis* sp.), increment zones are only delineated by dark deposed parenchyma bands (Argentina)

Fig. 5.99 Increment zones in a stem section of witte parelhout (*Aspidosperma marcgrafianum*). The tree is from a site that regularly has two drought periods each year, Surinam

In contrast to temperate latitudes, trees of the deciduous rain forest form increment zones. The cambium builds xylem evenly without seasonal periodicity. Increment zone boundaries are mark only by density differences or narrow parenchyma bands (Fig. 5.98).

In the subtropics and tropics with distinct rainy and dry seasons, as well as on trees with periodic leaf fall, cambial growth is temporarily interrupted (Sachsse 1991). This is marked by increment zones with distinct color variations or different frequency and arrangement of the conductive vessels (Fig. 5.99). These increment zones cannot be used to determine the age of the trees because trees can create one or more increment zones of varying widths depending on the yearly fluctuations in rainfall and drought.

Recent studies on the dating of tropical trees were conducted on *Cedrela odorata* and seven other tree species in the Amazon basin. They correlated the concentration of the oxygen isotope ($\delta^{18}0$) and of the carbon isotope ($\delta^{13}C$) in the increment zones with total annual precipitation levels. The result showed that growth sections are still not securely datable by 20 years. Reasonably safe values were only derived from growth periods of approximately 40 years (Brienen et al. 2012, Brienen 2013).

5.3.1.2 Causes

Tree ring formation is linked to the periodic change between the summer growing season and winter dormancy. As a result, trees in warm and humid tropical climates or in areas with rainy and dry periods only form increment zones. Typically, juvenile wood near the pith has wider tree rings. Tree rings become narrower as the tree grows wider and older and as competition in the canopy increases. Extreme weather events, calamities, and changes in stand density following heavy thinning can lead to irregular ring formation (Fig. 5.100). Solitary stems, often subjected to shifts in climatic stress, commonly exhibit variations in width within a tree ring as well as between different tree rings. Suppressed trees are often not able to form a (complete) tree ring (Rubner 1910). "Double" rings can form if warm fall temperatures stimulate new cell formation after the initial growth period has ended.

Instead of annual tree rings, tropical trees in deciduous rain forests or areas with extended periods of rainfall and drought form increment zones. These vary significantly depending on the weather conditions during the year, competition with neighboring trees, but also due to genetic predisposition (see Fig. 5.99).

See Plate 5.18 for tree ring variations in softwoods photos and Plate 5.19 for ring porous and diffused porous hardwoods from temperate zones and the tropics photos.

Fig. 5.100 The influence of thinning on the annual tree ring growth of an 80-year-old spruce *(Picea abies)*

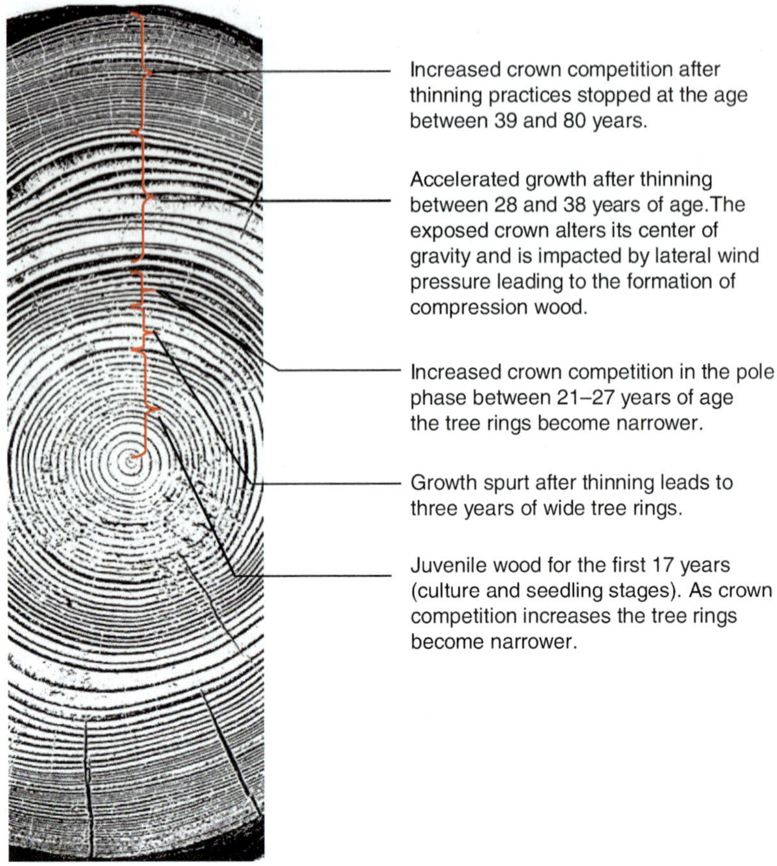

Chronicle of a 48 year-old spruce (*Picea abies*). Thuringian forest, Germany:

- 2005 felled
- 2003 drought year
- First thinning 1994
- Snow damage in 1980 in sapling stage
- Planted 1957

(Photo: G.-E. Weber)

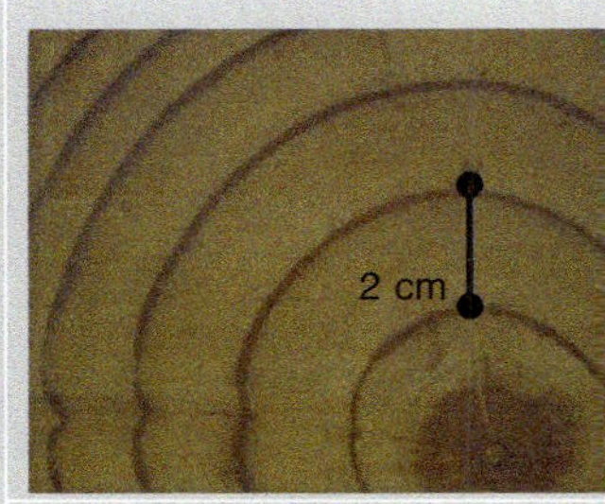

Monterey pine (*Pinus radiata*)

with very wide tree rings.

Dwarf juniper (*Juniperus communis, var. nana*),

82 tree rings,

Ø 0.16 mm/JR, Hamningberg, Barent Sea, Norway.

Tree rings in softwoods influenced by site and forest management measures

Pine (*Pinus sylvestris*) with wide rings narrowing to barely visible as it ages

Douglas fir (*Pseudotsuga menziesii*) cross section with uniform tree rings

Giant red cedar (*Thuja plicata*)cross section with up to 500 tree rings from an old growth forest (British Columbia, Canada)

Wavy, irregular tree rings in an 80-year-old fir (*Abies alba*) (Thuringia Forest, Germany) (Photo: G.-E. Weber)

Plate 5.18 Tree ring variations in softwoods

Beech, diffused porous, cross section (*Fagus sylvatica*) with red heart (Germany)

Walnut, semiporous, cross section (*Juglans regia*) (Germany)

Ash, ring porous, cross section (*Fraxinus excelsior*) with brown heart (Germany)

Robinia, ring porous, (*Robinia pseudoacacia*) (Germany)

Burmese rosewood (*Dalbergia bariensis*) one of the few semi-ring-porous species of the tropics (Vietnam)

Ausubo, diffused porous, (*Manilkara bidentata*) cross section, irregular spiral growth seen in the varied colored, concentric rings (Surinam)

Purple heart, diffused porous, (*Peltogyne venosa*) cross section. Increment zones are barely visible (Surinam)

Zwart parelhout, diffused porous, (*Aspidosperma corrugata*), heavily pronounced increment zones (Surinam)

Plate 5.19 Ring porous and diffused porous hardwoods from temperate zones and the tropics

Fig. 5.101 Different wood densities in the narrow and wide tree rings of this spruce (*Picea abies*) lead to varying shrinkage behavior in the wood. The tension is neutralized by a ring and radial shakes

Fig. 5.102 25 narrow tree rings in the core of this 36-year-old pine (*Pinus*) indicate suppressed growth in sampling stage. Average growth of 30 cm/year produces an approximately 7.5 m long, fine-limbed stem section

5.3.1.3 Prevention

The practice of thinning forests "frequently, but in moderation" keeps stand spacing consistent and can have a positive influence on tree ring formation (Burschel and Huss 1997). In commercial forests, however, the contrary economic axiom "seldom, but intense" usually prevails. One thing is for certain: the greater the value increment of a stand, the more cautious the thinning (Mette 1984).

Commercial plantations space trees geometrically and target maintenance to reduce crown, water, and nutrient competition. This positively affects the formation of uniform annual tree rings or increment zones.

In old growth forests or secondary forests of the boreal zone as well as in the rain forests of the tropics, fluctuations in the annual tree ring widths or increment zones are ultimately always subject to the local conditions on site.

5.3.1.4 Impact on Use

Variations in tree ring width can cause cracks in standing trees and particularly in dry timber (Fig. 5.101). In dried boards, extreme variations between wide and narrow tree rings lead to an increase in cracks and warping.

The internal knottiness of a tree can be diagnosed from the sequence of its tree rings around the pith. Wide tree rings at the core indicate strong juvenile growth, generally associated with a significant amount of branches in the lower stem area. Because of this, some timber sorting standards stipulate that pine must have a radius of 5 cm around the pith with at least 12–17 tree rings to be classified as A grade. This requirement attempts to ensure that the maximum tree ring width of 3–5 mm is not surpassed. With other high-grade timber, explicit rules exclude wood from a radius of 5 cm around the pith in the quality grade assessment (Fig. 5.102). This wood is mainly used to produce veneer sheets and rolls (TGL 1977c, Landesforstverwaltung Baden-Württemberg, 2002, SSLEF 1997).

In rotary cut and sliced veneer, irregular tree ring sequences produce unique and distinctive wood patterns.

Contrary to earlier opinion, no discernible correlation exists between tree ring structures and the tone quality of string instruments (Baltrusch 2003; Ziegenhals 2009).

5.3.1.5 Technological Adaptation

Wood with wavy tree ring patterns and marked variations in ring width cannot be used for higher-quality purposes (veneer, cutting wood, decorative pieces). There are no restrictions for mechanical or chemical wood pulping for chip or fiber board, paper, and pulp.

Plate 5.20 shows examples of tree rings being put to excellent use.

Radial split shingle from a giant cedar (*Thuja plicata*) (British Columbia, Canada)

Chapel with shingles from a coastal redwood (*Sequoia sempervirens*) (Silverton, USA)

Radial split spruce shingle (*Picea abies*) 1800 m above sea level with signs of abrasion, approximately 80 years old (Bergell, Switzerland)

12-year-old shingles from walaba (*Eperua* sp.) wood. Only ca.10 % of the wood can be used for straight shingles; 90 % is unsuitable due to spiral grain (Paramaribo, Surinam)

Wooden farm animals cut from spruce (*Picea abies*). This unique woodworking technique practiced since 1810 in the town of Seiffen, Germany, uses branchless, narrow-ringed wood to produce figures which are then hand carved and painted (Kirsche 1994). The ring rotation technique is regarded as one of the first examples of mass production in the history of wood technology

Plate 5.20 Objects putting tree rings to special use (wooden shingles, wood turning, musical instruments)

Making cello parts from spruce wood (*Picea abies*) in the Bohemian Erz Mountains, Czech Republic

Guitar parts made of spruce (*Picea abies*) hung to dry for several years (Vogtland, Germany)

Violin parts made from pine scantlings (*Picea abies*) from Carlsfeld and the Fichtel Mountains stored for at least 20 years (violin manufacturer Sch egel) (Erlbach, Vogtland, Germany)

Narrow-ringed spruce (*Picea abies*) with little latewood intended for the back of a zither from the Erz Mountains, Vogtland, Germany

Plate 5.20 (continued)

Fig. 5.103 Dead, severely right spiraling "scrub pine" on a rock formation

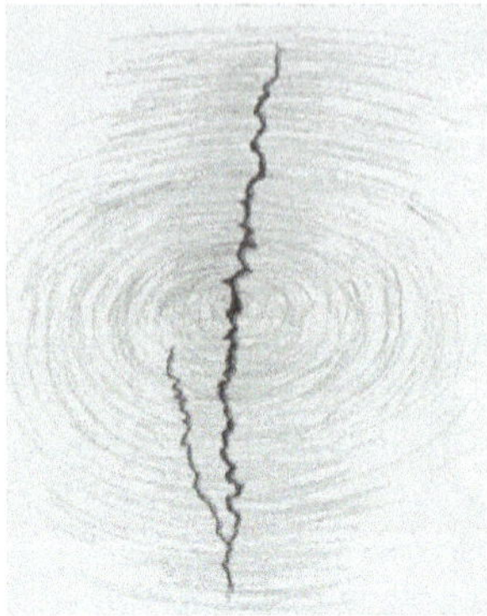

Fig. 5.104 Jagged initial cracks at face cut are the first signs of interlocked grain

5.3.2 Grain Orientation

5.3.2.1 Spiral Growth (Richter 2003a)

5.3.2.1.1 Description

Some trees grow in a spiral direction around the pith either to the right, from the lower left to the upper right (Fig. 5.103), or to the left, from the lower right to the upper left. Earlier literature defined right and left spirals as direct opposites (right clockwise, left counterclockwise) (Knuchel 1934).

Spiral grain is evident externally from a tree's bark, mouldings, and fluted stem sections. In debarked round wood, dry cracks on the surface are an indication of spiral grain. The direction of the spiral growth visible on the bark surface, however, does not necessarily reflect the direction of the spiral grain in youngest tree rings because the direction of the grain can switch long before it becomes visible on the bark (Knigge and Schulz 1966). Alternating spiral can be determined at felling when the initial face cut is made in the stem (Fig. 5.104). Spiral grain can be positively identified in the radial split piece of wedge wood.

Many species spiral in a preferred direction (Harris 1989, Schmid 1998):

Right spiral: Common in horse chestnut, beech, English oak, pear, holly, and juniper (*Aesculus hippocastanum, Fagus sylvatica, Quercus* ssp., *Pirus communis, Ilex aquifolium, Juniperus communis*).

Left spiral: common in gingko (*Gingko biloba*), the juniper species (*Juniperus rigida*), and plum (*Prunus domestica*) (Durst 1955). Left spiral is less common than right spiral.

Spiral with a single change in direction: Many conifers, especially spruce (*Picea abies*), exhibit left spiral near the pith and then change direction and spiral to the right as they age. Abrupt changes in grain orientation can be found at different stem heights as well as within a single tree ring (Schmelzer 1977; Wanninger 1989).

Interlocked spiral with periodical changes in direction: Many tropical trees, especially upper-story trees, periodically alternate the direction of their grain (Knigge and Schulz 1959), even between the early- and latewood of a increment zone (Thinley et al. 2005). This characteristic is often called interlocked grain.

Spiral grain appears, to a lesser or greater degree, in almost all older trees (Mayer-Wegelin 1956). Statistical studies by Kulikov showed that 78 % of the trees in West Siberian pine forests and 88 % of trees in the pine, larch, and mountain forests of the Urals have spiral grain (Durst 1955).

In general, spiral grain becomes more pronounced as a tree increases in size. According to research involving 4450 oak (*Quercus* ssp.) trees, only 20 % with a BHD of 10–20 cm spiraled to the right, compared with 60–70 % with a BHD exceeding 70 cm (Burger 1941).

Red alder rarely spirals. According to Neumeister, no spiral-grained birch (*Betula pendula*) has been identified (Gayer and Fabricius 1949). This, however, contradicts findings from barrel makers in the past who once used young birch trees, split lengthwise, to make barrel hoops. Records show they purposely selected birch trees from the center of stands because the wood was less twisted than trees from the stand perimeters frequently effected by wind exposure (Burckhardt 1879).

See Plate 5.21 for photos of images of spiral grain in trees from temperate regions.

Visible right spiral in a spruce (*Picea abies*) must be considered in grading

Giant fir (*Abies grandis*) with right spiral

Pine (*Pinus sylvestris*) with right spiral only visible on the dry, debarked stem

Western juniper (*Juniperus communis*) with severe left spiral on extreme site (Utah, USA)

Ponderosa pine (*Pinus ponderosa*) with severe right spiral on extreme site (Utah, USA)

Coastal redwood (*Sequoia sempervirens*) with left spiral (California, USA)

Old beech (*Fagus sylvatica*) with severe right spiral made visible by the mouldings

The frost cracks on this oak (*Quercus robur*) follow the left spiral grain of the stem

Young ash (*Fraxinus excelsior*) with pronounced right spiral

Willow (*Salix* ssp.) with extreme left spira

Hornbeam (*Carpinus betulus*) with left spiral

Chestnut (*Castanea sativa*) with left spiral (Loire, France)

Extremely fluted cherry (*Prunus avium*), with right spiral (Liechtenstein)

Plate 5.21 Images of spiral grain in trees from temperate regions

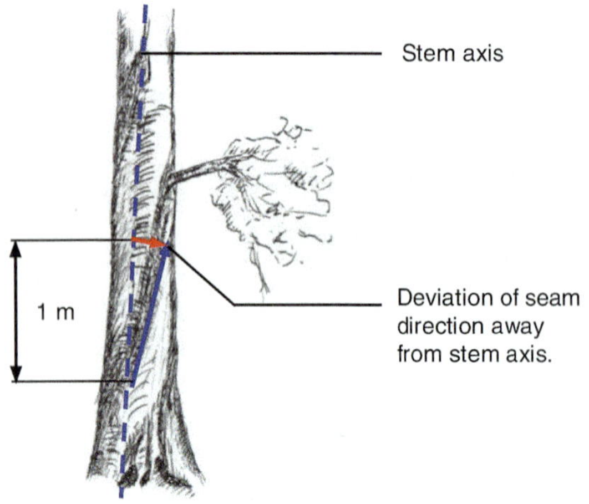

Fig. 5.105 Measuring spiral growth on a beech (*Fagus*) with mouldings

Measurement: According to HKS, the deviation of a tree's bark away from the stem axis is measured in inches per running meter, rounded down to the nearest centimeter (Landesforstverwaltung Baden-Württemberg 2002)

The CEN (DIN 1997a) specifies the deviation of the bark from the stem axis, measured on the spot of the greatest grain slope. With butt logs, the first meter is not included in the measurement because of its undulating grain at the stem base. The RVR (RVR 2014) follows the reasoning of the CEN (Fig. 5.105). Spiral grain spruce and fir should be measured only when the spiral is visible on the bark surface (Weidner et al. 1999). Weaker spiral can, however, only be observed on the dry cracks of debarked stems that have dried below the fiber saturation point.

See Plate 5.22 for spiral grain assessments in timber sorting photos.

The direction of the exposed inner bark in this veneer oak (*Quercus* ssp.) indicates right spiral grain	Right spiral oak (*Quercus* ssp.), on a submission as hardwood flooring 107 €/m³ sold –∅-proceeds from submission: Ei = 360 €/m³	Beech (*Fagus sylvatica*) with right spiral visible on the fluted trunk, quality grade C	Tension cracks along the left spiral grain degrade the wood to fire wood (*Fagus sylvatica*)	Severe left spiral spruce (*Picea abies*) log splits open in a spiral. No longer suitable for sawn timber, quality grade IFK

Plate 5.22 Spiral grain assessments in timber sorting (trees from temperate zones)

5.3.2.1.2 Causes

Probable causes of spiral growth are genetic predisposition, type of cambial cell formation (Fig. 5.106), age, site conditions, and prevailing wind effects (Burger 1941; Harris 1989; Hartig 1901; Krahl-Urban 1953).

From a tree's perspective, spiral grain means greater bending strength, *turgor* pressure, and *tensile strength* in response to predominately unilateral pressure. As a result, spiral grain increases the tree's chances of survival (Mattheck 1997; Schweingruber and Isler 1991; Steuer 1990). Thus, for example, in Lapland, pines spiral almost exclusively to the right the closer they grow to the timberline. This is also evident on exposed nutrient-poor sites in the rocky sandstone region of the Elbe River in Germany. The extremely twisted pine trees growing there are also called "spiral pines."

Trees with widespread crowns, especially fruit trees, need to have a particularly high tensile and bending strength to withstand wind pressure. Spiral grain works like a coiling rope which turns in the direction of the wind and becomes ridged.

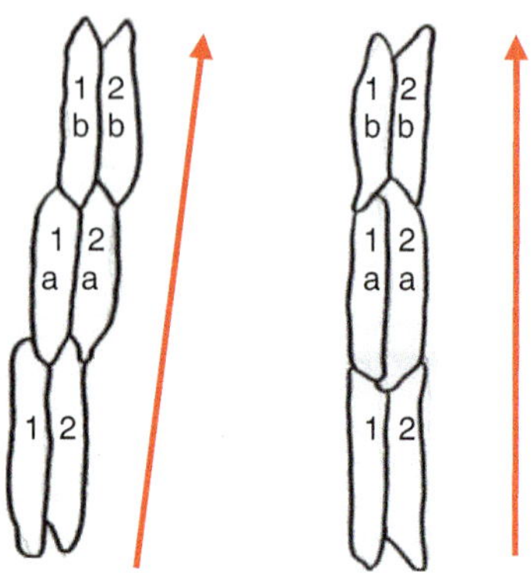

Fig. 5.106 Anatomical causes, parallel cambial cell division (*left*) leading to spiral growth, and opposed cell division leading to straight growth (*right*). (Cell division starts with cell 1; cell 2b is the last cell formed.) (In reference to Harris (According to Harris 1989))

In tropical trees, spiral grain works much like a reinforcement, stabilizing the trees with narrow stems and crowns that compete for sunlight (see Sect. 3.3). In an exploratory study of 33 randomly selected wooden boards each from different South American tropical species, 18 exhibited spiral growth. Tropical trees without spiral growth (e.g., wiswiskwari (*Vochysia guianensis*)) often crack after being felled due to the release of tension, because the interlocking effect of spiral grain is missing in the radial direction (Niemeier 2013).

Dominant trees without spiral grain, such as old spruce and larch, often exhibit shear stress cracks in the lower stem sections, particularly on windy sites. The stem base cracks under the tension on the windward side and under compression stress on the leeward side. This is rarely observed in spiraling trees because the spiral grain neutralizes the shear and pressure stress (Mayer-Wegelin 1956). A spiral stem results in greater bending strength, compression, and torsion stability by predominately unilateral pressure (Richter 2008b) (Fig. 5.107).

Calculations show that at a grain slope of 15–30°, a tree's maximum pressure tolerance increases by about 1.5 %, because the pressure is distributed equally in a longitudinal and lateral direction (Richter and Hennig 2001). It is possible that this effect could, over the course of evolution, give spiral trees an advantage by increasing their chances of survival (Mattheck 1997).

Spiral grain has an even greater impact on wood's tensile strength. Studies on wood cylinders with non-spiraling and interlocked grain showed that the bending strength of the interlocked grain cylinders was more than double (223 %) than that of the cylinder with the axis parallel grain orientation (Hansen 2004). These results corresponded with earlier findings from Thunell (1951), showing the benefits of spiral grain in stem and branch wood.

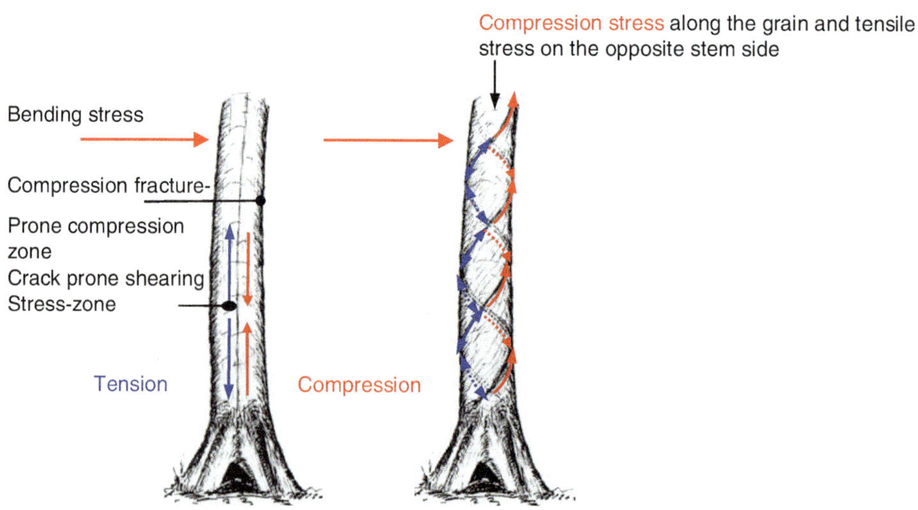

Fig. 5.107 Effect of shear pressure under bending stress in tree without spiral growth (*left*) and with spiral growth (*right*)

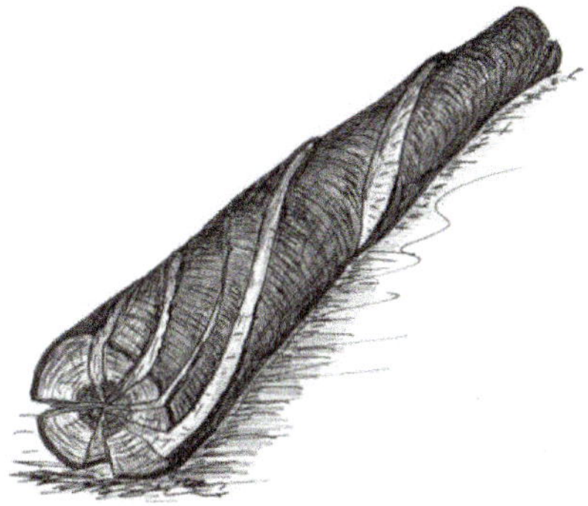

Fig. 5.108 Growth cracks in a beech (*Fagus*) along the spiral grain

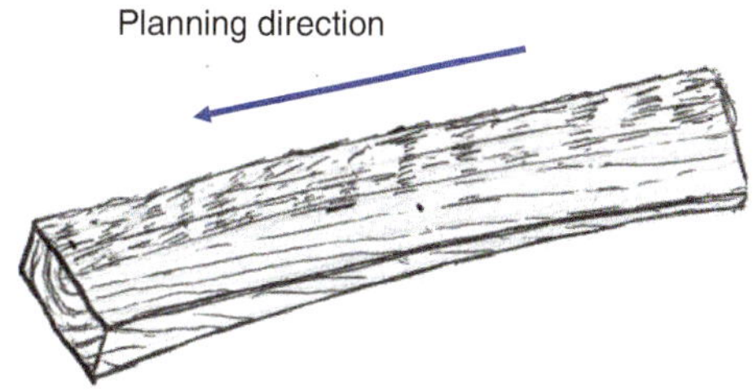

Fig. 5.110 Alternating surface quality on a planed board from a right spiral stem. Rough surface caused by planing against the grain

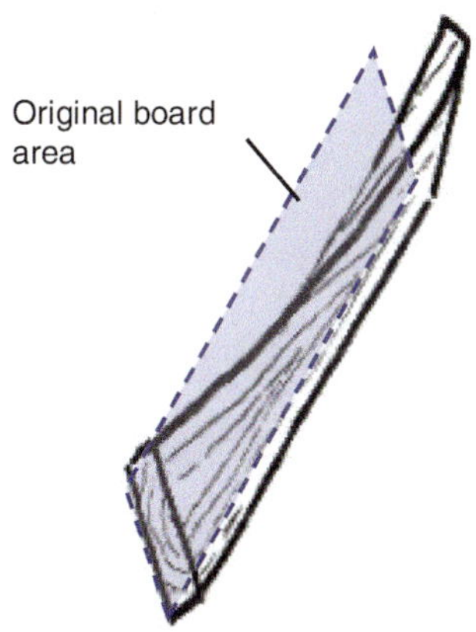

Fig. 5.109 A board from a right spiral stem warped during seasoning

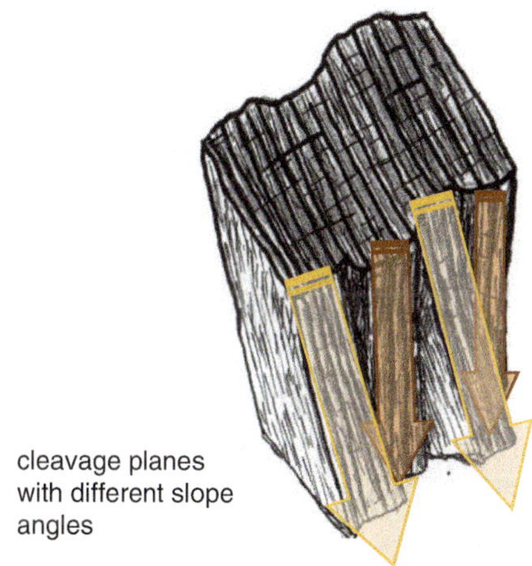

Fig. 5.111 Uneven cleavage planes for spiral-grained tropical wood

5.3.2.1.3 Prevention

In commercial forests, spiral trees should be removed as part of sound stand maintenance practices.

Spiral grain is undesired in crop trees.

Because spiral grain increases with age, all spiral grain trees should be harvested before natural regeneration occurs to prevent any economically adverse genetic properties from transferring to the next generation (Burger 1941). Old, well-maintained birch stands, therefore, do not have the same number of spiral grain trees as a stand with normal distribution.

An established seed stock stand should not contain any spiral grain trees (oak (Quercus ssp.), beech (*Fagus sylvatica*)). Individual spiral grain trees should not be used for seed collection.

The common occurrence of interlocked grain in wood from primary and secondary forests must be accepted.

5.3.2.1.4 Impact on Use

Spiral grain logs dried below a fiber saturation level of 30 % can crack along the spiral (Fig. 5.108). These logs are not suitable for lumber, but the wood can still be used for pulp and particle board.

Beams with spiral grain can be used in buildings, but only if they are not fixed to the structure. Sawn timber from spiral grain stems tends to warp in the direction of the grain (Fig. 5.109).

Sawn timber looses strength the more the grain direction deviates from the parallel. A slope in grain of 15° can result in a 40 % decline in bending strength (Niemz 1993).

Planning against the spiral grain produces a rough surface and split fiber (Fig. 5.110).

Wooden shingle makers avoid using spiral wood. They identify unsuitable wood by radially splitting a small piece of the stem. If the split wood is twisted, then the wood cannot be used as a shingle wood (Fig. 5.111).

See Plate 5.23 for impact of spiral grain on sawn timber photos.

Testing a spruce (*Picea abies*) stem for spiral based on split log trim	Larch beam (*Larix decidua*) with left spiral. The crack was already in the stem as a pitch pocket	The missing fiber significantly reduces the strength of the left spiral spruce (*Picea abies*) beam	Right spiral spruce (*Picea abies*) board after 100 years of dry storage	Warping evident in a left spiral beam in a Swedish log house after 200 years

Plate 5.23 Impact of spiral grain on sawn timber

Table 5.10 Effect of spiral on quality class (RVR 2014, beech and oak; DIN 2008a, b, c, softwood)

	Quality class			
Species	A	B	C	D
Spruce/fir (*Picea/Abies*)	≤3 cm/m	≤7 cm/m	≤10 cm/m	Unlimited
Pine (*Pinus*)				
Larch (*Larix*)				
Douglas fir (*Pseudotsuga*)				
Beech (*Fagus*)	≤2 cm/m	≤6 cm/m to 4. Stkl.	Unlimited	Unlimited
Oak (*Quercus*)		≤7 cm/m to 5. Stkl.		

On the other hand, in tropical wood, spiral grain has an aesthetical value due to the unique figure it produces in tropical veneer wood. Scheiber (1965) only assesses interlocking spiral as a defect if the grain orientation on the stem surface twists more than 30 % to the stem axis. For impact of spiral grain on the grade and quality class of various timber assessments, see Table 5.10.

See Plate 5.24 for identifying (interlocked) spiral in tropical timber photos and Plate 5.25 for interlocking spiral grain in tropical veneer wood photos.

5.3.2.1.5 Technological Adaptation

In the production of sliced veneer, saws are used which allow the veneer block to be positioned at the blade so a single cut can be made diagonal to the direction of the fiber. Cut timber must be stacked immediately using piling sticks and extra pressure added if needed so that the boards do not warp as they season.

In the manufacturing of planed lumber, it is necessary to ensure that the wood is planed in the direction of the grain.

The adverse effect of slope of grain on sawn timber strength can be identified with the help of machine grading (Pope et al. 2005).

The principle of spiral grain has been proposed as a practical model for constructing statically stressed buildings such as towers, silos, and large pipes (Richter and Hennig 2001).

See Plate 5.26 for spiral growth used as a model in construction.

Left spiral visible on the surface of a kromantikopi stem (*Aspidosperma* sp.)

This jagged crack in zwarte kabbes (*Diplotropis purpurea*) indicates internal spiral grain

Mawsikwari stem (*Erisma uncinatum*) with interlocked grain

Light and dark rings in this pisi (*Ocotea* sp.) cross section conceal interlocked grain (*left*). Radial section reveals extremely interlocked grain (*center*). Stem section split down the middle reveals grain inalternating directions (*right*)

Thin, alternating layers of interlocked wood in a zwarte kabbes (*Diplotropis purpurea*)

Wana (*Ocotea rubra*) with coarse interlocked grain (*left*). Split tangential the wood is smooth. Only fibers running diagonal are visible (*center*). Cracked plank on the right with uniform wood layers and varied grain directions

Growth stress in a wiswiskwari (*Vochysia guianensis*) leads the stem to crack when felled because the interlocking effect of alternating grain is missing

Gele kabbes board (*Terminalia dichotoma*) planed using dull blades–the fibers against the grain are erect

Plate 5.24 Identifying (interlocked) spiral in tropical timber (Surinam)

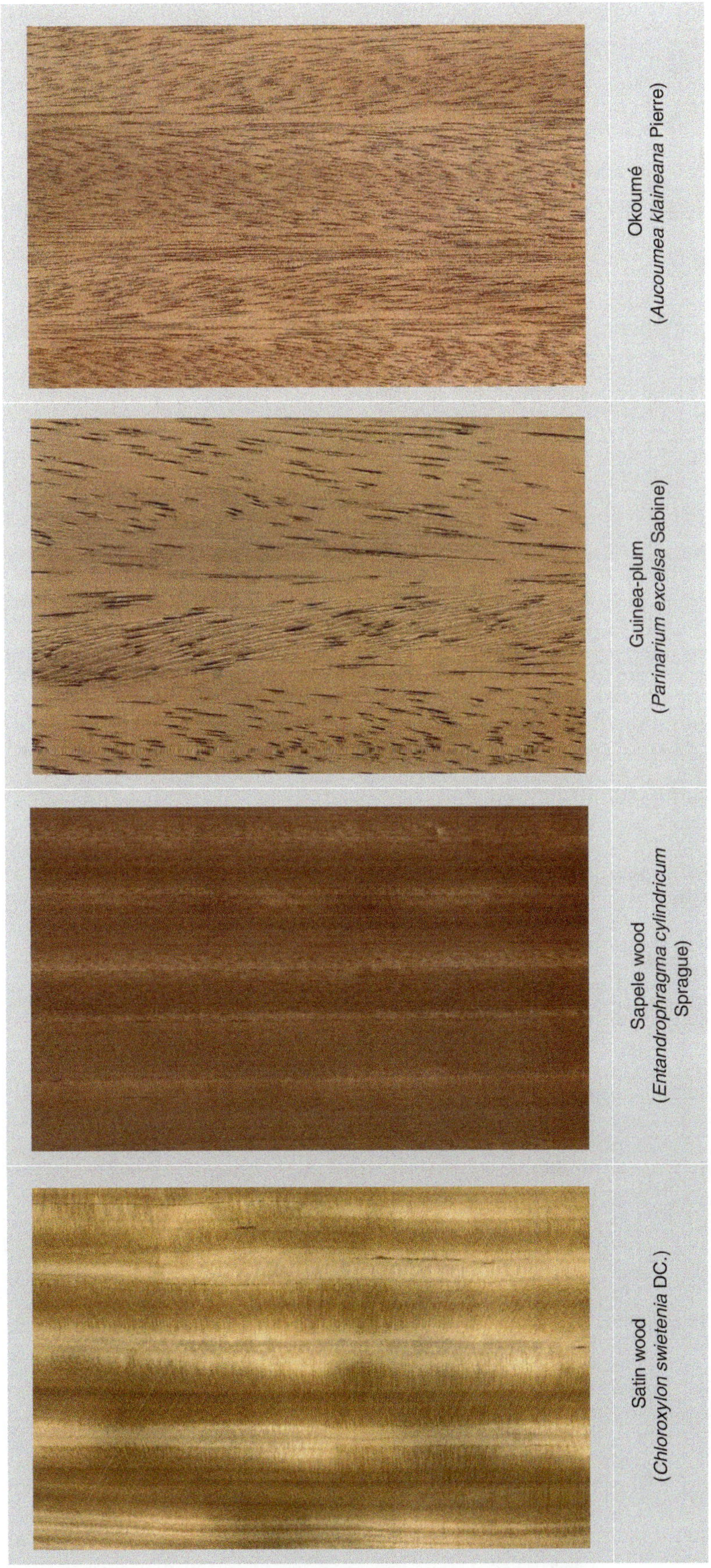

Plate 5.25 Interlocking spiral grain in tropical veneer wood

1. Field observation

Neighboring spruce (*Picea*) stems break in 125 km/h storms.

The stem with fiber running parallel to the stem axis breaks lengthwise near the base (*top*)

The spiral stem only breaks at 4 m height (*right*)

2. Building a function prototype

Functional principle of a spiral tower made of wooden slats constructed in three layers (Hennig and Flechsig 2000),

Wooden slats with slanted middle slat for tongue and groove.

Gage/brace

Spirally arranged steps enable a separate accent and decline.

Zinc plating between steps and tower casing

3. Preparation of test models

Left, test model with fiber running parallel to the stem axis ($Ø = 20$ cm, $H = 100$ cm);

Center, right spiral; and

Right, alternating spiral, inner layer right spiral outer layer left spiral (Photo: Hansen 2004)

4. Trial and evaluation

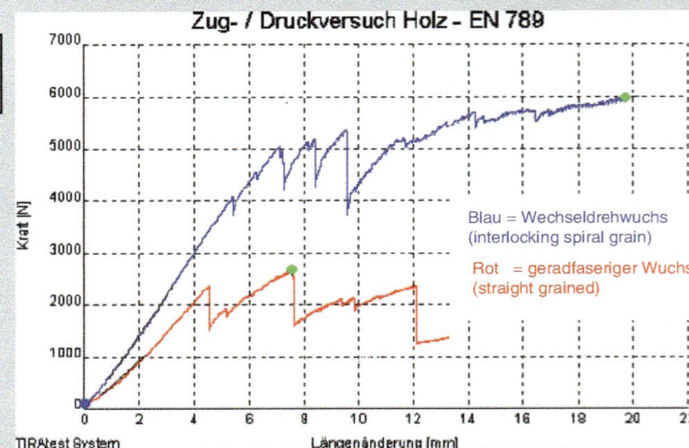

Bending test of braced cylinder test specimens 20 cm in diameter and 6 mm thick; force displacement diagram (Hansen 2004)

Plate 5.26 Spiral growth used as a model in construction

Spiral grain vividly illustrates how varied opinions can be when made from different perspectives, such as a wood processor's, a bionic scientist's, or an aesthetically oriented dendrologist's.

5.3.2.2 Grain Orientation: Curly, Fiddleback, and Hazel Growth (Richter 2004b)

5.3.2.2.1 Description

Wavy grain orientation or tree ring patterns generally occur in a tree's bole or buttress, running either transverse or perpendicular to the axis, and in tropical trees are often associated with interlocked grain. Depending on the direction of the waves, the grain is called either curly, fiddleback, or hazel growth. Since many authors confusingly use "curly" as a generic term for the more specific terms longitudinal and transverse curl, curly grain, fiddleback, hazel growth, and "bird step," the following attempts to offer a sufficiently precise definition of each term:

Curly grain, also called "elephant skin" or "washboard" effect, refers to wavy lateral contortions in the tree rings running transverse to the stem axis and visible on the stem surface. The wood fiber undulates with the tree rings (Fig. 5.112, red arrow).

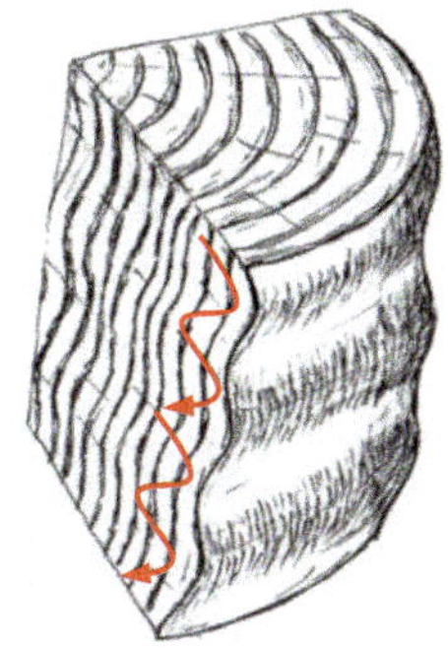

Fig. 5.112 Curly grain beech log (*Fagus sylvatica*)

Curly grain is relatively common in beech (*Fagus sylvatica)* and sycamore (*Platanus hispanica*) and less so in maple (*Acer* ssp.), horse chestnut (*Aesculus hippocastanum*), chestnut (*Castanea sativa*), pear (*Pyrus pyraster*), linden (*Tilia* sp.), birch (*Betula* ssp.), aspen (*Populus tremula*), oak (*Quercus* ssp.*)*, spruce (*Picea abies*), and larch (*Larix decidua*) (Mette 1984). In beech *(Fagus sylvatica*) and sycamore (*Platanus hispanica)*, the wide waves are also called "elephant skin" and the shorter, buckled waves at the buttress and under the branch collars in maples (*Acer* ssp.) are called "washboard" effect. In tropical trees, curly grain is sometimes found in combination with interlocked grain creating a moiré or feather pattern.

See Plate 5.27 for curly grain photos.

Curly grain beech (*Fagus sylvatica*), on the leeward of the prevailing wind (Thuringia Forest, Germany, 700 m elevation)

Oak (*Quercus* ssp.) with bumps typical of curly grain (Unterharz, Germany)

Curly grain can occur in all wood types; old grapevine (*Vitis vinifera*) (Madeira, Spain)

Basralocus (*Dicorynia guianensis*), cause of curly grain unknown (Surinam)

Hoogland mataki (*Symphonia globulifera*), curly grain likely influenced by microorganisms (Surinam)

Curly grain cambium compressions on the lower side of a heavy sycamore branch (*Platanus hispanica*)

Curly grain (compressed bark) at the buttress of a western red cedar (*Thuja plicata*), caused by alternating directions of cambium growth

Curly grain ash veneer (*Fraxinus excelsior*)

Curly grain chestnut veneer (*Aesculus octandra*) (Appalachia, USA)

Curly gain beech (*Fagus sylvatica*) veneer used for a table top

Plate 5.27 Curly grain

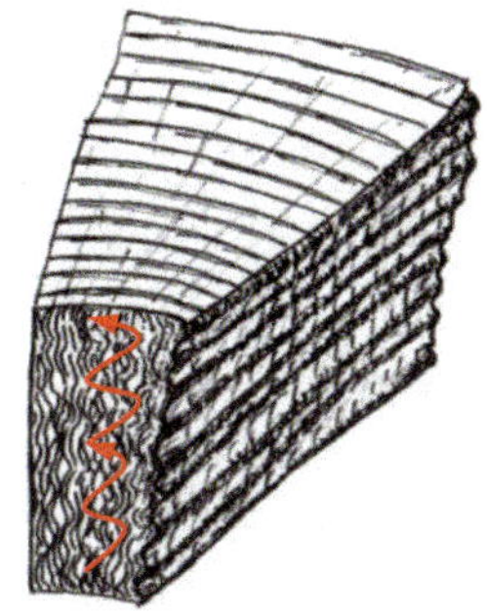

Fig. 5.113 Fiddleback on a split of sycamore maple log (*Acer pseudoplatanus*)

Fiddleback (e.g., in certain maples (*Acer* ssp.)) refers to wavy grain that runs transverse to the axis of the stem in a radial direction and is not visible on the stem surface. The wood fibers undulate in the tree rings (Fig. 5.113, red arrow). Fiddleback can only be determined on live trees by removing a piece of dry bark and identifying fine wavy marks left behind on the inside of the bark. Fiddleback is readily apparent on the radial face cut before felling. It creates a "washboard" effect in the split radial section. In a radial cut, the fibers are intersected at different angles. This leads to an optically interesting pattern of light and dark stripes (Hoadley 1990).

Fiddleback is also found in douka (*Tieghemella africana*), makore (*Tieghemella heckelii*), limba (*Terminalia superba*), and ash (*Fraxinus excelsior*) (Wagenführ 2007). Studies in Surinam examined 300 stored logs of various species from primary forests and found fiddleback figure in 20 % (!) of the bergigronfolo (*Qualea rosea*) and 5 % of the bolletrie (*Manilkara bidentata*). Fiddleback figure is sometime found combined with curly grain to create a flame pattern, with interlocked grain in a moiré pattern or a combination of the two (Niemeier 2013).

See Plate 5.28 for fiddleback photos.

The surface of this sycamore maple (*Acer pseudoplatanus*) trunk at 750 m elevation suggests highly figured wood (Vogtland, Germany)

Drying cracks in a hoogland gronfolo (*Ruizterania albiflora*) log reveal fiddleback grain (Surinam)

Fiddleback in this redwood (*Sequoia sempervirens*) after decades of weathering (California, USA)

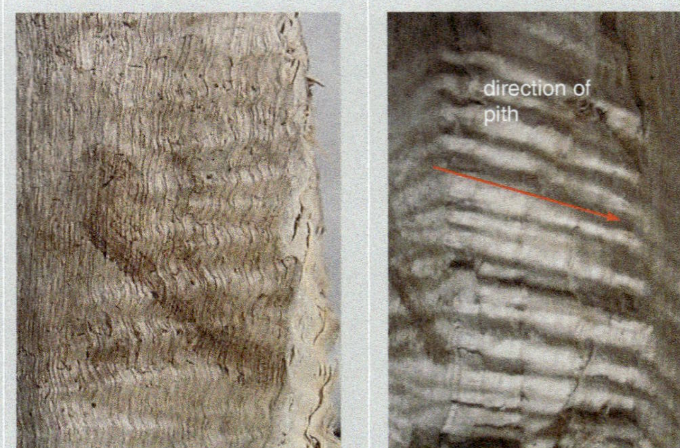

Small dry cracks reveal fiddleback on the debarked surface of a split piece of walnut wood (*Juglans regia*) (lle.). The same walnut piece in radial view (ri.)

Radial split fiddleback sycamore maple (*Acer pseudoplatanus*) used for a violin bridge

Fiddleback sycamore maple veneer (*Acer pseudoplatanus*) used for a guitar back

Fiddleback violin frame from sycamore maple (*Acer pseudoplatanus*)

Fiddleback akazie (*Acacia melanoxylon*) veneer

Fiddleback yellow buckeye (*Aesculus octandra*) veneer

Plate 5.28 Fiddleback

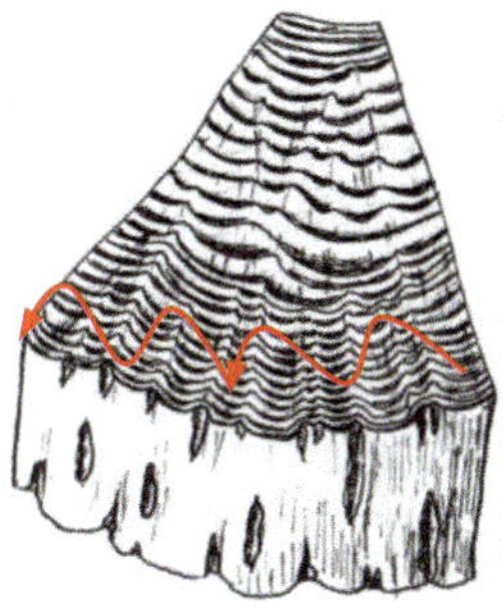

Fig. 5.114 Wavy grain and tree ring orientation on a hazel spruce (*Picea abies*)

Fig. 5.115 Spruce with "bird step" figure, radial cut, macroscopic

Hazel growth in spruce (*Picea abies*) and sometimes fir (*Abies alba*) refers to wavy, undulating grain or indented tree rings running parallel to the stem axis in a radial direction (Fig. 5.114, red arrow). The only rather unreliable external feature of hazel growth is a series of longitudinal dents in the bark. In debarked wood, hazel growth is visible in the elongated dents in the stem surface. The term "hazel growth" comes from its similarity with the anatomical structure found in hazel wood cross sections. The contorted grain visible in the radial section and the markings on the surface often resemble wings or tracks and are therefore sometimes called "bird step" figure (Fig. 5.115). Hazel growth spruce can occur equally in all phenotypes, also in comb, brush, and plate spruce. The likelihood of hazel growth increases with altitude (about 800–1,500 m) (Rüegsegger 1962).

Microscopically, hazel growth has a 40–50 % higher amount of pith rays compared to normal spruce (*Picea abies*) wood (Körner 1961). Tangentially, but especially in a radial direction, the wood exhibits a highly irregular tracheid distribution and orientation (Fig. 5.116). Cell density is more than 50 % greater than in normal spruce. Wood density is about 20 % above normal spruce and the strength values are higher (Rüegsegger 1962).

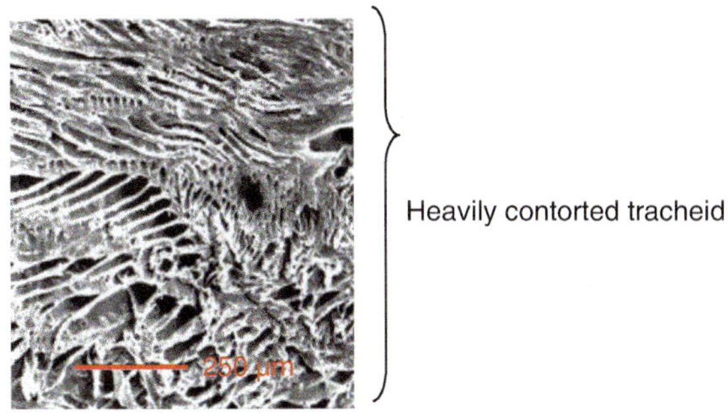

Fig. 5.116 Abnormal cell arrangement for spruce, radial cut (SEM photo: E. Bäucker)

Table 5.11 Synonyms for spruce and fir with hazel growth (Körner 1961)

Vogtland	Hazel spruce,
	Tonholz, Hahnentritt, Vogeltritt
German Alps	Hasel-, Stein-, Weißfichte
Bernese highlands	Hagelholz, Tonholz, Mändliholz, Ageholz, "glismets" (knitted) wood, equal : vogeltrittig, katzentrittig or hagelschlächtig
Switzerland	Schindeltanne (shingle fir)
Vorarlberg Bosnia	Zottelfichte (clumps spruce)
Bohemian Forest	Geflammtes Resonanzholz (flamed resonance wood)
	Zargenholz (frame wood)

Hazel growth wood has different names depending on the region. This reflects the varied appearances, properties, or intended usages (Table 5.11).

Hazel growth is also found in alerce (*Fitzroya cupressoides*), yew (*Taxus baccata*), as well as mansonia (*Mansonia altissima*), ash (*Fraxinus excelsior*) (Wagenführ 2007), basralocus (*Dicorynia guianensis*) and limba (*Terminalia superba*).

See Plate 5.29 hazel growth photos.

Debarked spruce stem with hazel growth (*Picea abies*) from 750 m elevation (Vogtland, Germany)

Hazel growth spruce cross section (*Picea abies*)

Hazel growth resonant wood processed into a sounding board for a zither (Ulrike Meinel, Fa. Adolf Meinel, Markneukirchen, Vogtland, Germany)

Resonant wood with "bird step" figure in a harpsichord, radial cut

Tangential split hazel growth spruce board. (*Picea abies*)

Weathered barn siding, Bernese Oberland, Switzerland, 1,900 m elev, about 50 years old, made from so-called "shingle fir" (*Abies alba*)

Plate 5.29 Hazel growth

Assessment: In commercial grading (HKS 2002a) and in the framework agreement for timber trade in Germany (RVR 2014), the terms curly grain, fiddleback, and hazel growth are not included. The term curly grain is only defined in the European CEN as "*tight, regular transverse bands appearing on the wood surface with a wavy fiber orientation*" (DIN 2001b, p. 7). Thus, these wood characteristics should be assessed using the general grading standards set in HKS (2002a) and graded according to RVR (2014) or CEN (DIN 1997c, 1998b). Curly grain is usually considered a wood defect because the contorted tree ring sequences can impede wood surface processing and the wavy tree rings only have a limited decorative effect.

In contrast to curly grain, fiddleback and hazel growth are highly valued. When suitably manufactured, fiddleback and hazel growth figure combined with spiral grain can significantly improve the practical value of the wood.

5.3.2.2.2 Causes

Curly grain is generally considered to have physiological causes. Bosshard (1984) writes about "functional tropism," in which the function of the cells formed in the cambium outweighs their structural purpose. In this case, the water supply system is affected for the benefit of the storage system (increase in wood ray parenchyma). In the affected regions, this leads to reduced radial growth.

Mette (1984) suggested curly figure occurs as a result of unilateral stress to the stem which increases with age. Another cause could also be the greater elasticity associated with the wavy grain orientation. This assumption is supported by observations on beach growing on a wind exposed slope at 700 m elevation. There, curly grain always appears on the leeward side of the trunk. A genetic predisposition may lead wind influences to stimulate growth and produce curly grain.

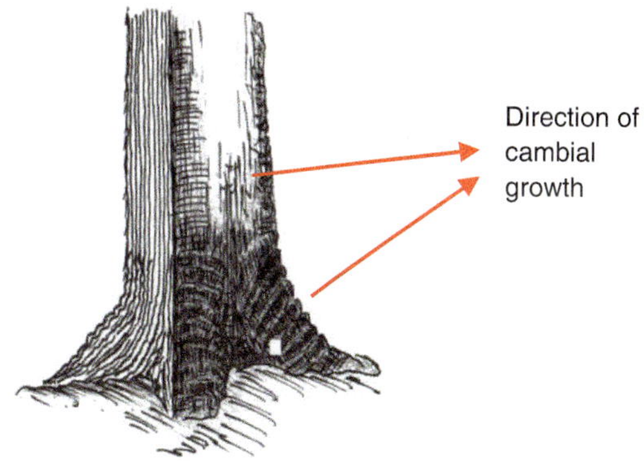

Fig. 5.117 Bark and cambium compression (curly grain) on a maple (*Acer*) buttress

Schweingruber and Isler (1991) also suspect cambium compression in stem curvatures and branch collars (Fig. 5.117).

Genetic disposition is widely believed to be the cause of *fiddleback* figure. Growth stimulation is also suspected, which could be caused by wind-induced stress.

As cause for *hazel growth* figure, Bosshard (1984), Mette (1984), and Rüegsegger (1962) all point to a localized shortage of cambium in the area of wood rays (similar to fluting). Greyerz (1919) claims the increase in pith rays to be the main cause of the anomalous growth.

The causes for the particularly wavy grain in "bird step" spruce are unknown (Schweingruber and Isler 1991). The term "hail spruce" reflects the assumption that heavy hail storms are the cause of cambium injuries that bring about hazel growth.

Curly, fiddleback, and hazel growth figure are the most common types of grain patterns. Combined with interlocked grain, they can produce strikingly beautiful wood. Table 5.12 shows several of these combinations.

See Plate 5.30 Different deviations in fiber orientation photos.

Table 5.12 Possible wood textures from combinations of different deviations in fiber orientation

[a]Preferred cutting directions: R = radial, T = tangential, C = cross section. *Yellow fields*: basic deviations in the fiber orientation. *White fields*: combinations of different fiber orientations

Grain deviations	Interlocked	Curly	Fiddleback	Hazel growth
Interlocked	**Striped** R[a]  Sapele-mahagony (*Entandrophragma cylindricum*)  Afrormosia (*Pericopsis elata*)	**"shelled" T[a]**  maple (*Acer pseudo-platanus*) **pommelé T**  Australian white birch (*Schizomeria ovata*)	**frost patter R** (swirling)  Sapele-mahagony (*Entandrophragma cylindricum*)	**Dents** on the stem surface: figure running longitudinal to stem axis **T** 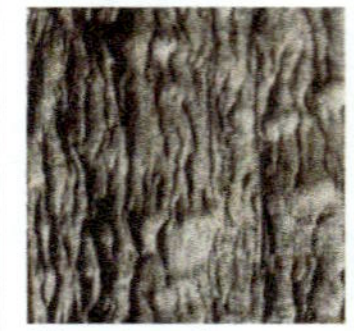 Wanakwari (*Vochysia tomentosa*) 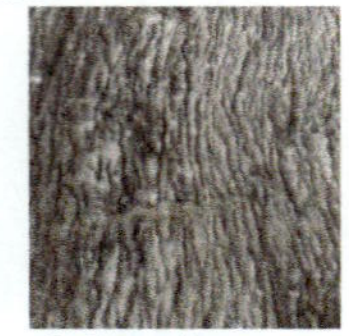 Salikwari (*Tetragastris ssp.*)
Curly	**Moiré** (slanted stripes)  Olive (*Olea europaea*) **Feather R**  Australian white birch (*Schizomeria ovata*)	**Curly** **T**  Beech (*Fagus silvatica*)  Yellow buckeye (*Aesculus octandra*)		
Fiddleback	**Moiré** (slanted stripes) **R**  Bolletrie (*Manilkara bidentata*)	**Flamed** **T** 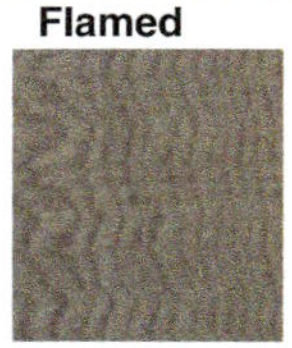 Redwood (*Sequoia sempervirens*)  Oak (*Quercus crysolepsis*)	**Fiddleback** **R**  Maple (*Acer pseudo-platanus*)  Yellow buckeye (*Aesculus octandra*)	
Hazel growth (bird step)				**Hazel** **R C[a] T**  Spruce (*Picea abies*) **C**  Limba (*Terminalia superba*)

Live sycamore maple (*Acer pseudoplatanus*) with spiral, slanted, crisscrossing curly, and fiddleback grain, likely with valuable figured wood (Erz Mountains, Germany)

Valuable maple veneer (*Acer pseudoplatanus*) from a combination of spiral, curvy, and fiddleback grain

Heavily barked birch suggests fiddleback grain (Northern Sweden)

Birch cup (*Betula pendula*) with curly and fiddleback grain (Finland)

Figured Douglas fir (*Pseudotsuga menziesii*) from a combination of curly and fiddleback grain

Moiré pattern in the olive (*Olea europaea*) created by combination of interlocked and curly grain (*left*), while in the bolletrie (*Manilkara bidentata*), it is a combination of interlocked and fiddleback grain (*right*).

This gronfolo (*Qualea rosea*) appears to have interestingly figured wood

Pommelé pattern in a sapele (*Entandrophragma cylindricum,*) created by a combination of interlocked, curly, and fiddleback grain

Plate 5.30 Different deviations in fiber orientation

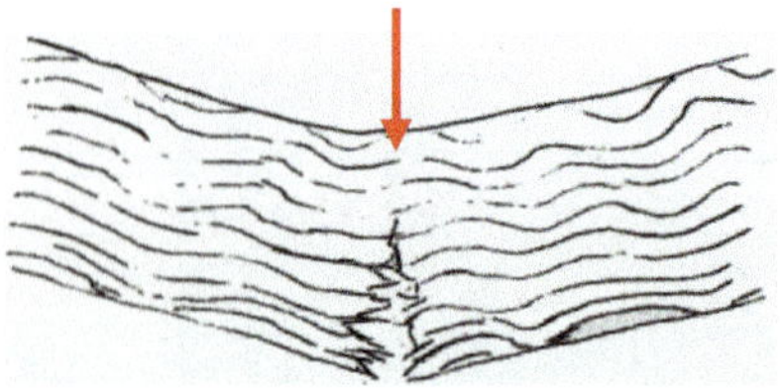

Fig. 5.118 Reaction to bending stress by curly grain wood

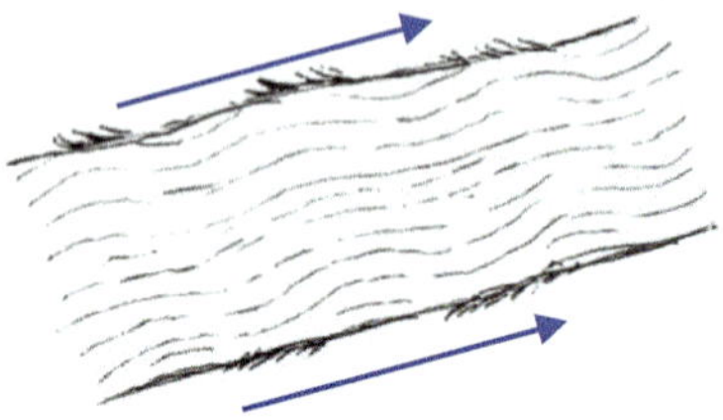

Fig. 5.119 Rough curly grain surface after planing with a blunt saw

5.3.2.2.3 Prevention

Curly, fiddleback, and hazel growth figure cannot be influenced. Affected trees should be removed if they negatively impact the intended use. They should be maintained if they can be used for a certain purpose.

In hardly any other characteristic group does the intended purpose determine so significantly whether the characteristic is seen as an unpleasant defect or a special feature that greatly increases the wood's value. For the most part, too little attention is paid during timber selection to "unconventional" grain and tree ring orientation.

5.3.2.2.4 Impact on Use

The strength properties of sawn lumber, especially slats and laths, decline in wavy grain wood (Fig. 5.118). Planing against the wavy grain produces a rough surface (Fig. 5.119). The wood becomes difficult to split.

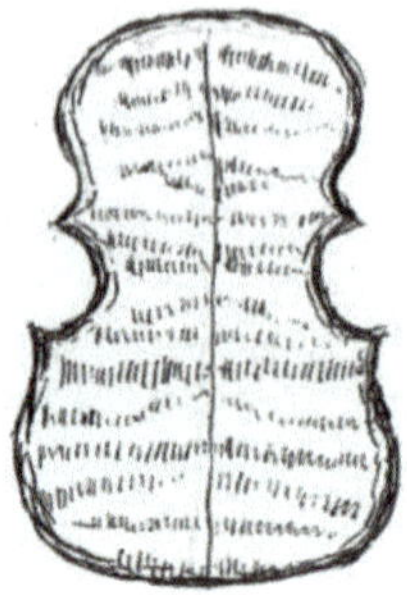

Fig. 5.120 Fiddleback violin back

The HKS of Thuringia (Thüringer Landesforstdirektion 1997) discusses the disadvantages but also the benefits of wavy grain: for example, curly grain in beech (*Fagus sylvatica*) veneer wood is unacceptable because the wavy grain makes the veneer and plywood hard to straighten and easy to tear. On the other hand, fiddleback is highly valued because the wood's decorative effects make it valuable as face veneer and as soundboards or frames for musical instruments (violin and guitar) (Fig. 5.120). The same applies for the rare hazel growth spruce, in particular "bird step" figure is prized for making violin and guitar backs, harpsichords, etc. (Fig. 5.115). Many woodworkers claim that the wavy grain has superior acoustical properties. The relatively heavy wood sounds brighter than usual narrow ring spruce with little latewood from mountain regions.

5.3.2.2.5 Technological Adaptation

Well-sharpened tools ensure a smooth surface also in small areas with contorted wood fibers.

Special attention should be given not to overbend fiddleback and hazel growth wood used for veneer and musical instruments.

For spruce (*Picea abies*) and fir (*Abies alba*) trees, the term "shingle fir" reflects a preference for using dense hazel growth spruce as shingle wood.

6 Biotically Induced Wood Characteristics

6.1 Impact of Microorganisms

6.1.1 Necroses

6.1.1.1 Description

Necroses or cancers are callused areas of localized dead cambium tissue, most commonly found in hardwoods, mainly in beech (*Fagus sylvatica*), but also in oak (*Quercus* ssp.) and maple (*Acer* ssp.) (Fig. 6.1). During the vegetation period, a dark slimy mixture of sap and microorganisms briefly oozes from the wound, giving the disease the name "slime flux." In cross sections, these wounds appear as T-shaped markings (Fig. 6.2).

Necroses also occur in trees from the subtropics and tropics. The high incidence of infection often results in large areas of dead bark.

See Plate 6.1 for necrotic wounds photos.

Fig. 6.1 Beech (*Fagus sylvatica*) stem with severe necrosis

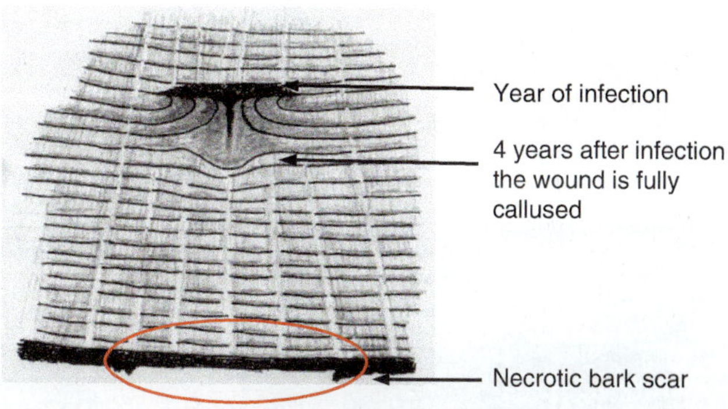

Fig. 6.2 Infection and callusing creating a T-mark

C. Richter, *Wood Characteristics: Description, Causes, Prevention, Impact on Use and Technological Adaptation*, DOI 10.1007/978-3-319-07422-1_6, © Springer International Publishing Switzerland 2015

The necrosis on this beech (*Fagus sylvatica*) stem has degenerated into cancer

Elongated necrotic wound on a beech (*Fagus sylvatica*) (*left*) and stem cross section with T-mark (ingrown dead bark), condition about 15 years after attack.

Section of a beech (*Fagus sylvatica*) board with necrosis.

Beech (*Fagus sylvatica*) stem with old bark scar indicating a solid, deep necrotic wound

Strong necrosis (cancer) in an ash (*Fraxinus excelsior*)

T-mark in a maple (*Acer pseudoplatanus*)

T-mark in an oak (*Quercus* ssp.)

Bergigronfolo (*Qualea rosea*) large necrotic area without cambial growth

Tingimonisali (*Tetragastris hostmannii*) with severe slime flux

Plate 6.1 Necrotic wounds

6.1.1.2 Causes

Bark necroses often occur after periods of extreme winter cold spells, long periods of drought, or following bacterial and fungal infection (Table 6.1) (Altenkirch et al. 2002). Necroses in trees from all climate zones are probably best understood as the result of a chain of effects. Prerequisite for the infection are small (microscopic) bark injuries which serve as entryways for microorganisms. In most cases, insects are the main culprits. For example, the woolly beech scale (*Cryptococcus fagisuga*) bores into the bark, creating portals for *Nectria* in beech trees (*Fagus sylvatica*), while another set of scale insects infests oaks (*Quercus* ssp.).

A proven link also exists between bark beetle (*Trypodendron domesticum*) infestation and *Nectria ditissima* infection (Grüner and Metzler 2003).

The microorganisms damage or destroy the cambium cells. The tree protects itself from infection by building callus tissue and walling off the wounded area (Altenkirch et al. 2002; Bosshard 1984a; Knigge and Schulz 1962). As the infection spreads, the risk that the injury will degenerate into cancer increases.

Measurement: Necroses are major wood defects. They can only be accurately identified in stem cross sections. Their position in the stem can be approximated as follows (Table 6.2, Fig. 6.3):

6.1.1.3 Prevention

Necroses are difficult to avoid. Either too little is known about their cause to prevent them or it would require a disproportionate amount of effort to protect against the smallest infection.

Necrotic trees should be removed as part of sustainable stand maintenance (Fig. 6.4) because they are carriers for further infestation. Since necrotic wood does not grow, there is no increase in the amount of wood.

6.1.1.4 Impact on Use

Necrotic wounds are undesirable in all types of wood manufacturing because the markings visible on the bark do not accurately reveal the depth of the injury, the amount of wood discoloring, the degree of insect infestation, and the extent of callusing in the damaged part of the stem. Necrotic wood can no longer be used as merchantable timber (Fig. 6.5).

Knigge and Schulz (1962) categorize beech (*Fagus sylvatica*) wood into four quality levels depending on the size and duration of the disease:

1. Healthy, uniformly pale yellow to pink colored wood
2. Dark-colored protective wood with compartimentalization (tylosis) under the dead cambium, gray colored when dried

Table 6.1 Causes of necrosis (selection)

Tree species	Responsible microbes and indirect influences
Beech (*Fagus sylvatica*) (Grüner and Metzler 2003; Wujciak 1976)	Beech cancer (*Nectria coccinea*) (*Nectria ditissima*)
Oak (*Quercus* ssp.) (Mette 1984)	Oak beetle borer, oak scale insects (*Keramidae*)
Maple (*Acer* ssp.), poplar (Populus tremula), alder (*Alnus glutinosa*) (Mette 1984)	General: insect infestation and other external influences cause bark injury. Microorganisms penetrate the wounds and infect the cambium

Table 6.2 Estimating the depth of the necrosis (T-mark) based on the shape of the bark scar (Wujciak 1976)

Appearance	Size
Long scar with recognizable callus, scar height to scar width >3: 1 (line type)	<3 cm depth
Pronounced scar	<7 cm depth
Wide scar, scar height to scar width <3: 1 (round type)	>3 cm depth
No callus	>>7 cm depth

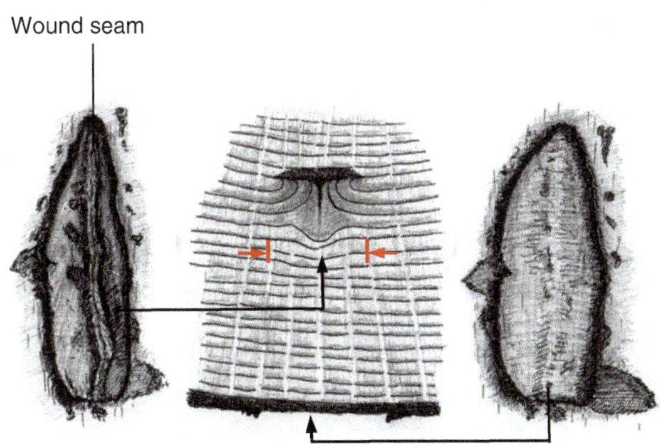

Fig. 6.3 The appearance of a bark scar on a beech (*Fagus sylvatica*) as a (weak) diagnostic feature for the wound depth of necrosis. *Left* 5 years after infection with overgrowth seam, *right* 20 years later

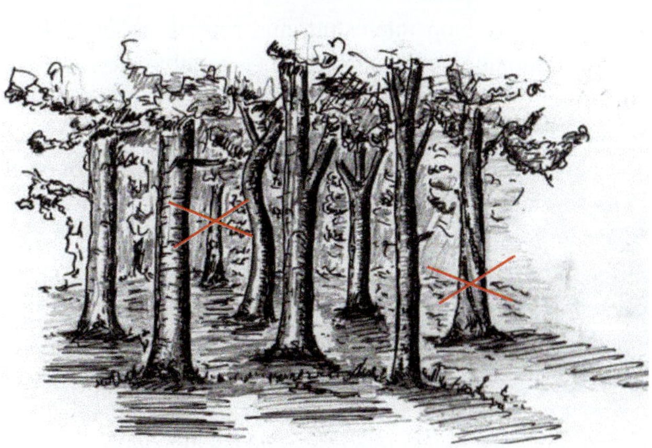

Fig. 6.4 Removal of beech (*Fagus sylvatica*) with necrosis (*red marking* indicates mature timber harvest)

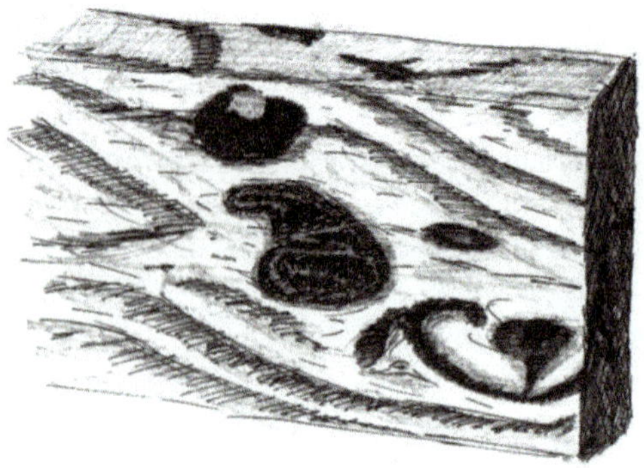

Fig. 6.5 Unusable beech (*Fagus sylvatica*) board with severe necrosis. Pieces of bark are ingrown

3. Discolored wood with heavy beetle damage (*Xyloterus domesticus*) or (*Hylecoetus dermestoides*)
4. Wood completely destroyed by white rot

Under "limited slime flux" Frommhold (2001) means maximum 10/lfm. High-quality wood (A and B) with even just a few necrotic wounds (2–3/lfm) will be degraded to a lower quality class.

6.1.1.5 Technological Adaptation

The bark scar can be used as a (weak) diagnostic feature for determining the depth of the necrosis in the stem (Fig. 6.1).

Veneer sheets with necrotic areas must be either trimmed off or centered in the middle of the plywood sheet.

Necrotic wood has areas with tylosis. Therefore, it cannot be fully impregnated and is, for example, no longer suitable for use as railroad ties.

6.1.2 Growth Anomalies Caused by Growth-Stimulating Microorganisms: Galls, Burls, and Witches' Brooms (Richter 2005b, 2006a)

6.1.2.1 Description

Trees often develop interesting and sometimes quite large growth anomalies on their stems. Generally referred to as galls or burls, they can either have a relatively smooth, clear surface or be covered in shoots or twigs.

Many trees also exhibit unique broom-like clusters of small branches in their crowns, colorfully called witches' brooms.

While most galls and burls look very similar, both clearly differ in appearance from witches' brooms. Studies on the causes of such growth abnormalities have yet to establish conclusive results. The growths are often referred to as tumors since they each involve (pathological) tissue development with excessive cell growth that is obviously no longer governed by the tree's original genetic code. They are sub-defined as follows:

Galls are cell proliferations resulting from an infection of the cambium or individual buds in the tree's juvenile stage.

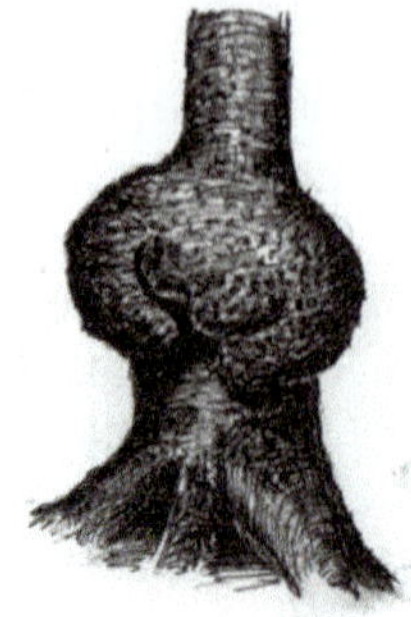

Fig. 6.6 Spruce (*Picea abies*) gall

Fig. 6.7 Birch (*Betula pendula*) burl

Tree rings or increment zones in galls are much wider than in normally developed stem wood. In benign tumors, the wood appears generally healthy. In malignant tumors the microbial degradation often leads to wood necrosis. Galls are relatively common among spruce (*Picea abies*) (Fig. 6.6) and birch (*Betula pendula*), but also found among hardwoods and many other softwood and hardwood species from the temperate and tropical zones. A rough exterior is caused by excessive bark development and not by epicormic shoots!

See Plate 6.2 for malignant tumors photos and Plate 6.3 for galls (benign tumor) photos.

Burls result when buds continuously develop into epicormic shoots in the stem bark or root area, also called bud proliferation (Fig. 6.7). The shoots regularly die away leaving behind attractive patterns in the wood grain. Burls are common in walnut (*Juglans regia*), poplar (*Populus* ssp.), sugar maple (*Acer saccharum*), and linden (*Tilia* ssp.). In (sub)tropical trees, they are common in madrones (*Arbutus menziesii*), afzelia (*Afzelia* ssp.), myrtle (*Myrtus communis*), and padouk (*Pterocarpus* ssp.).

See Plate 6.4 for burls photos.

Witches' brooms are characterized as early sprouting, negatively geotropic buds and shoots on branches (Fig. 6.8). They are particularly vital and can assume a significant part of a tree's assimilation (birch (*Betula pendula*), pine (*Pinus sylvestris*), etc.).

See Plate 6.5 for witches' broom photos.

Fir (*Abies alba*) cancer (Black Forest, Germany) (*left*); fir (*Abies alba*) cancer, (Carpatho-Ukraine) (*right*), cause: *Melampsorella caryophyllacearum*

Spruce (*Picea abies*) cancer (fir bark disease), single cancer (*left*), and many cancers (*right*), cause: *Nectria fuckeliana* (Oberharz, Germany)

Young larch (*Larix*) cancer, cause: *Lachnellula willkommii*

Cancer in a Murray pine (*Pinus murrayana*) (Columbia Mountains, Canada) (*left*), and cancer in a late-blooming black cherry (*Prunus serotina*), (Tharandt, Germany) (*right*). Tissue damage during dormancy leads to wound craters, causes: fungi, similar larch cancer (*Lachnellula willkommii*)

Beech (*Fagus sylvatica*) cancer (*left*). An interruption in assimilate flow towards the root lead to increased wound callusing. Infested stem (*right*) (Tharandt, Germany), cause: *Nectria ditissima, Nectria coccinea*

Partially malignant tumor on a wiswiskwari (Vochysia guianensis) (Surinam), cause: unknown

Plate 6.2 Malignant tumors

Spruce gall (Tharandt, Germany), cause: unknown

Pine gall (Lapland, Finland), cause: unknown

Gall intergrowth between two redwoods (*Sequoia sempervirens*) (California, USA), cause: unknown

Galls in western white pine (*Pinus monticola*) (British Columbia, Canada), cause: unknown

Douglas fir bark disease caused gall-like bumps on *Pseudotsuga menziesii* (British Columbia, Canada), cause: unknown

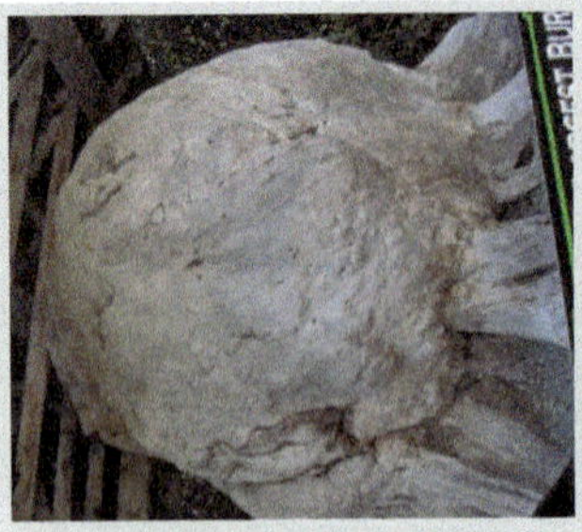

World's largest gall from Western hemlock (*Tsuga heterophylla*), diameter ca. 3.5 m, (Port Mc. Neill, Vancouver Island, Canada)

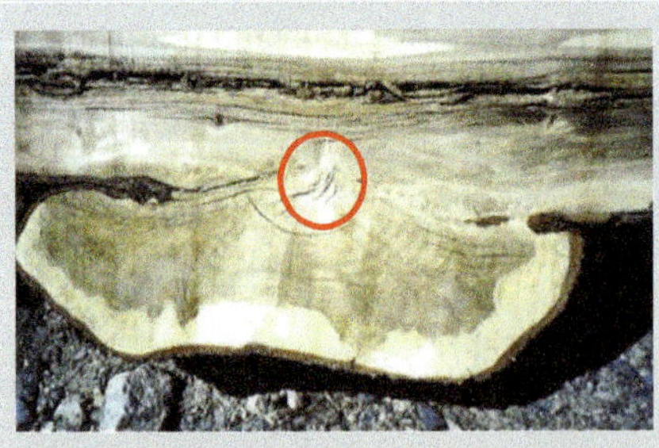

Old birch (*Betula*) gall (*left*). Right younger birch (*Betula*) gall, cut open, showing initial point of cancer, causes: Nectria fungi.

Beech (*Fagus sylvatica*), gall (*left*), cut (*right*) showing initial point of cancer, causes: *Nectria galligena* or *Nectria ditissima*, *Nectria coccinea*

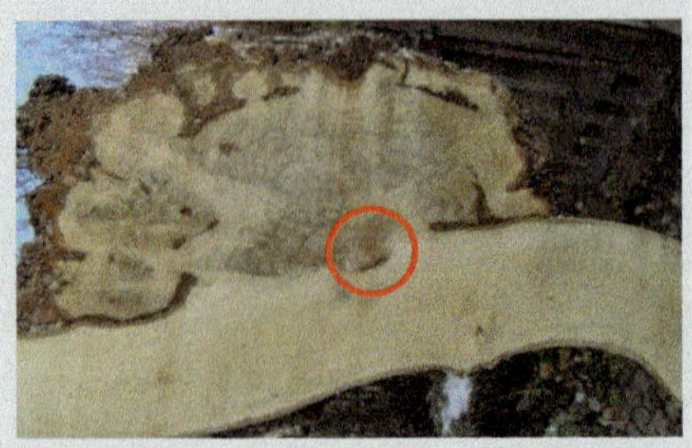

Cancer on a hornbeam (*Carpinus betulus*) (*left*). *Right* the same cancer cut open showing initial point of cancer (Saxony, Germany), cause: *Nectria* fungi

Plate 6.3 Galls (benign tumor)

Ash gall (*Fraxinus excelsior*), cause: *Nectria*-fungi/bacterial ash cancer, genera *Xanthomonas*, *Pseudomonas*

Red alder(*Alnus glutinosa)* with galls (Tharandt, Germany), cause: parasitic fungi

Oak gall (*Quercus* ssp.) (Brandenburg,Germany), cause: oak aphid *Lachnus roburis*/ *Stereum rugosum*

Gall cn basralocus (*Dicorynia guianensis*)covered with epiphytes (Surinam)

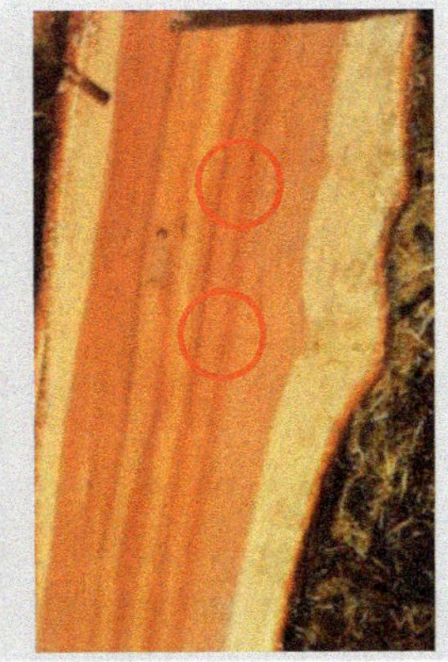

Gall-like bumps on hoogland mataki *(Symphonia globulifera*) with initial point of cancer growth (right) (Surinam)

Plate 6.3 (continued)

Left sugar maple (*Acer saccharum*) (Kensington Park, London), GB;, *right* birds eye veneer from sugar maple, cause: parasitic fungi

Mahogany burl (*Swietenia mahagoni*) (West Indies), cause: unknown

Burl on a mountain bergigronfolo (*Qualea rosea*) stem (Surinam)

Left: Elm burl veneer (*Ulmus glabra*), *right*: walnut burl veneer (*Juglans regia*)

Water sprout burl (anxiety shoots) on an intermediate maple (*Acer pseudoplatanus*) (*left*) and oak (*Quercus robur*) (*right*) with *physiological causes*. They occur when latent buds sprout as a result of light deficiency (suppressed crown) or due to sudden increase in light supply (canopy clearing)

Left, linden (*Tilia cordata*) burls with bud proliferation (Greenwich, GB).
Right, oak (*Quercus ssp.*) with bud proliferation (Gran Canaria, Spain), cause: probably injury, insect infestation or other external influence, infection by microorganisms

Young burls on birch (*Betula pendula*) (*left*) and old birch burl (*right*), (Germany), causes: probably injuries, insect infestation and other external influences, infection by microorganisms

Plate 6.4 Burls

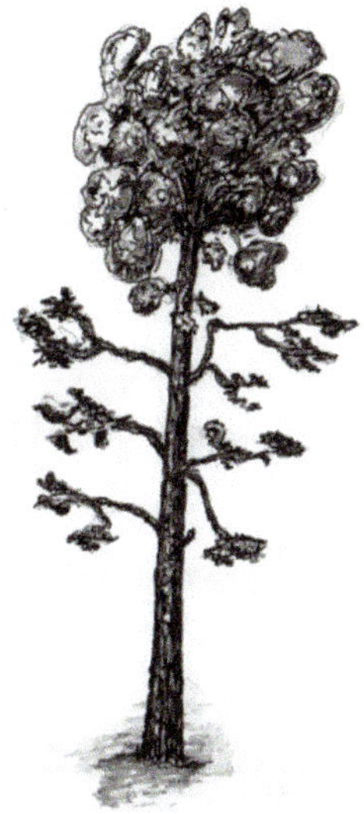

Fig. 6.8 Pine (*Pinus sylvestris*) with witches' broom (Finnmark, Norway)

Necroses assume an intermediate position, depending on whether they heal by scarring or whether they develop into cancer.

Measurement: Some timber grading systems classify *galls* and *burls* as abnormal stem contours. Diameters, therefore, are measured at an equal distance above and below the abnormality. In certain cases, the excrescences can be omitted from calculations of overall wood volume. Basically, trees with galls or burls are graded as unsuitable for sawn lumber and are either left standing or are only sold as D-quality industrial wood (SSLEF 1997). If the gall or burl wood is to be sold for veneer, this should take place only after the last and highest bid in a stumpage sale. Since it is difficult to determine wood volume, high-quality burl wood is sold by weight (Matzek 1983).

Witches' brooms grow on twigs and branches in a tree's crown. They are not suitable as lumber and are therefore not measured.

6.1.2.2 Causes

Tree galls and burls occur more frequently in cities, parks, and the open countryside than in forests. These trees are subject to a greater risk of injury and considerable stress (emissions, dryness, heat). In addition, cities and more frequented areas present a higher rate of infection.

When buds and bark are injured by forestry operations, fire (western red cedar), insects, or small growth cracks, the wounds often become entryways for parasites. Certain viruses, bacteria, and fungi are then able to stimulate tissue growth in order to create favorable living conditions for themselves (Altenkirch et al. 2002; Gottwald 1983). In the infected areas (cambium, buds), nutrients or growth hormones are increasingly enriched. The subsequent excessive radial growth causes pathological excrescences to develop that can develop into tumors (Wagenführ and Scheiber 1989). They can be benign tumors, as in spruce (*Picea abies*) or poplar *(Populus* ssp.*)* burls, or malignant as, for example, in fir, larch, or beech cancers *(Abies alba, Larix decidua, Fagus sylvatica)* or witches' broom.

Malignant tumors can develop from frost damage as, for example, in larch cancer (*Larix decidua*). The damage is magnified by the ongoing assault from microorganisms during the vegetation period (Schwerdtfeger 1981). The constant shift between attempting to seal off the wound, and the subsequent death of the latest layer of cambium tissue, often brings about oddly shaped tumors.

Attention! The above pathological growth anomalies have nothing to do with clusters of new shoots that often sprout up after canopy openings (increased light) or canopy closures (anxiety suckers) or due to repeated removal of water sprouts (as often the case with linden trees (*Tilia* ssp.) lining a street) (Wagenführ and Scheiber 1989). In these cases, physiological processes cause buds to grow from the meristem around branch scars.

Mites and insects can stimulate growth to their benefit, resulting in growth anomalies in the tree trunk and branches. Mites are direct causes of witches' broom in birch (*Betula pendula*) (Georgi 1965), and wasps are identified as causing various leaf galls. According to studies by Eschrich (quoted in Schwerdtfeger 1981), the oak aphid (*Lachnus roburis*) attacks the most recent shoots of oak *(Quercus* ssp.*)* trees at the nursery stage. The diseased cambium cells repeatedly burst open and are then sealed over. These cells have probably been genetically "reprogrammed," because the gall continues to grow.

The infection of a young sprout or bud may explain why the starting point of a gall growth is always located on a branch collar in the lower, juvenile section of the stem. Schwerdtfeger (1981) proposed that the cause of gall formation in pine (*Pinus sylvestris*) and spruce (*Picea abies*) is the same as in oak *(Quercus* ssp.*)*. In spruce (*Picea abies*), moths (*Dichelia histrionana*) can create openings for infection in the side branches of young stems. The mechanisms of infections are often unclear.

Studies on three tropical tree species, mountain bergigronfolo (*Qualea rosea*), hoogland gronfolo (*Qualea albiflora*), and mawsikwari (*Erisma uncinatum*), observed that insects (species unknown) permanently bored through the stem surface to the cambium. The trees then attempted to callus over the stem wound created by the bore hole. As the insects retreated for air, they created new wounds which the tree again tried to seal off. The results are calluses occurring sporadically or throughout the stem.

See Plate 6.6 for unique growths photos.

Witches' broom in a pine (*Pinus sylvestris*) (Finnmark, Norway), cause: *Phytoptus*-species

Witches' broom in a spruce (*Picea abies*) (Tharandt, Germany), cause: *Arceuthobium pusillum*

Black spruce (*Picea mariana*) with witches' broom. (Dease Lake, British Columbia, Canada), cause: unknown

Dead witches' broom on a Douglas fir (*Pseudotsuga menziesii*) (Wyoming, USA), cause: unknown

Witches' broom in a larch (*Larix sibirica*) (Khuvsgul Nuur, Mongolia), cause: unknown (Photo: M.Bürger)

Taphrina betulina causes burls and witches' brooms in this birch (*Betula* ssp.) (Bavaria, Germany).

Old witches' broom in a birch (*Betula pubescens*) (Finnmark, Norway), cause: *Ascomycete Taphrina betulina.*

Cluster of witches' brooms in lodgepole pine *(Pinus contorta)* caused by dwarf mistletoe (*Viscum* sp.) Right top: pine protects itself against injuries with pitch, right bottom: full grown dwarf mistletoe (Oregon, USA).

Witches' broom in a western juniper (*Juniperus communis*) caused by dwarfmistletoe (*Viscum* sp.), (Nevada, USA)

Plate 6.5 Witches' broom

Bark burl on a beech (*Fagus sylvatica*) (*left*), showing initial point of growth with the xylem of the stem (*middle*) and cut open lower side of gall with the initial point of growth at the top (*right*), cause: unknown

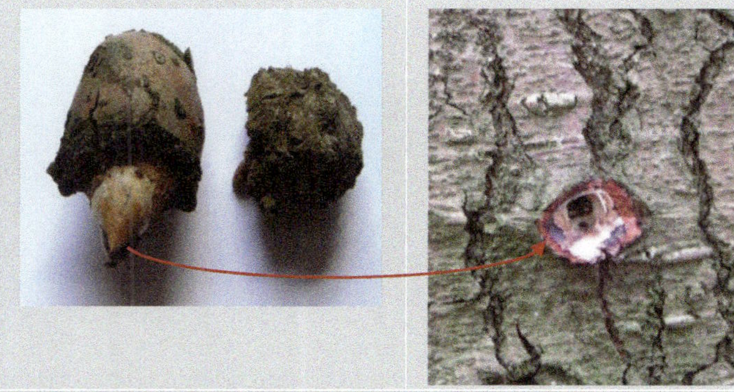

Bark burl from a Douglas fir (*Pseudotsuga menziesii*) (left) and point of growth with the stem xylem (right).

Tumor on a myrtle wood (*Laurus nobilis*), cause: unknown (Madeira, Portugal)

Eucalyptus (*Eucalyptus regnans*). With tumor (Madeira, Portugal), cause: unknown

Tumor on unknown tropical species (Surinam), cause: unknown

Tumors on poplar (*Populus* sp.) (*Germany*), cause: *Agrobacterium tumefaciens*

Tumors as artistic element, "Mother Nature," artist: Lindner, Daetz-Stiftung, Lichtenstein/Germany

Plate 6.6 Unique growths

Cylindrical bumps mountain gronfolo stem (*Qualea rosea*) (*left*). Cylindrical pitting to the top of the bump (*center*). The pitting extends to the stem core (*right*). The tree calluses over the bump with each new growth ring (Surinam) cause: unknown, probably insect

Bumpy protrusions on the stem surface of a mawsikwari (*Erisma uncinatum*). The tree calluses over the larva holes (*above*). Bumps on the stem surface of a hoogland gronfolo (Qualea albiflora), visible only after removing the bark (*below*). In both cases, the cause is an unknown insect, (Surinam)

Plate 6.6 (continued)

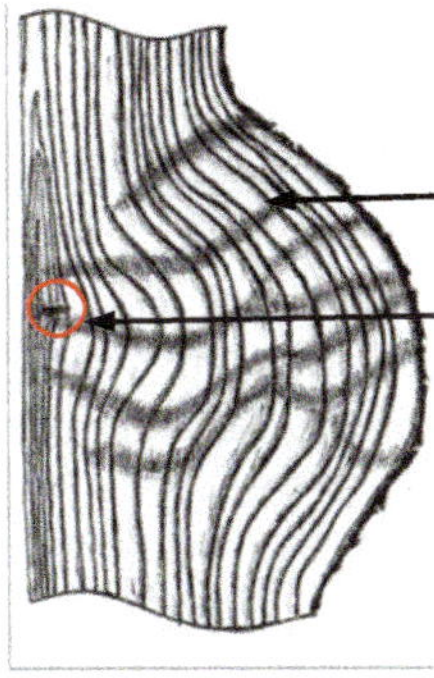

Fig. 6.9 Spruce (*Picea abies*) gall cross section

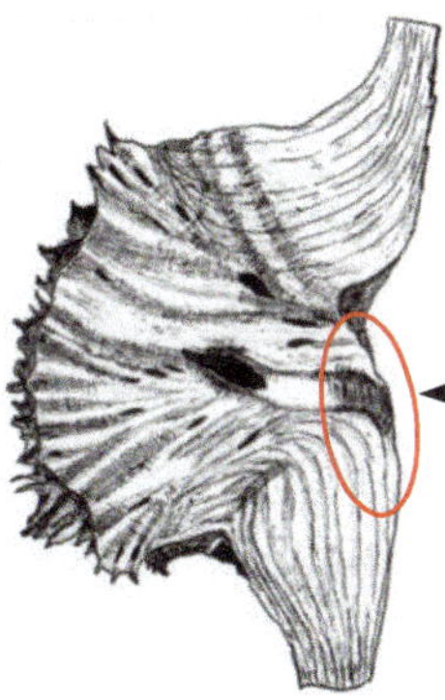

Fig. 6.10 Birch (*Betula pendula*) burl cross section

Also *mistletoe plants* (*Viscum* sp.), which are specific to certain host plants, can trigger the growth of witches' brooms. This was observed among North American pines (Pinus *contorta*, *P. flexilis*) and Douglas fir (*Pseudotsuga menziesii*) and Mediterranean cypress (*Cupressus sempervirens*) (see Sect. 6.3).

Cross sections of benign galls, burls, and witches' brooms have the following typical properties:

Galls: The excessive growth *starts from an initial point* early in the tree's life (Fig. 6.9). As the gall grows, the infected area remains free of latent buds or water sprouts. Ripple-like marks appear in the radial section.

Bark burls and ulcers are defined as special forms of galls:

- Bark burls are xylem formations in the bark of hardwoods such as beech (*Fagus sylvatica*), oak (*Quercus* ssp.), red oak (*Quercus rubra*), linden (*Tilia* ssp.), maple (*Acer* ssp.), alder (*Alnus glutinosa*), but also firs (*Abies alba*). The small spherical tubers on the stem are easy to remove (Butin 1996).
- Ulcers are tissue growths caused by bacteria, viruses, or genetic mutations (Butin 1996) in the cambial and bark parenchyma cells.

Burls: A cluster of shoots starts from *an initial areal on the stem mantle* (Fig. 6.10). These shoots regularly die off

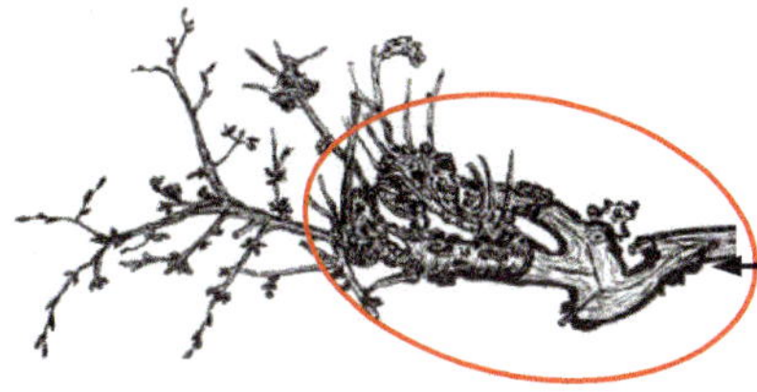

Fig. 6.11 Witches' broom in a birch (*Betula pendula*)

and are repeatedly sealed over with another growth layer. In tangential sections, they appear as an arrangement of circular points like contour lines.

Witches' broom: A *mass of buds and short shoots* is stimulated to form a thicker bushy growth – known as a witches' broom (Fig. 6.11). The affected twigs expand unevenly.

Galls, burls, and witches' brooms share the fact that they are initiated by microorganisms that cause damage to living tissue over entry portals (injuries). However, the microbial effects are different. They can be obvious pathogens (necroses, malignant tumors), but can also lead to a type of symbiosis with the host (galls, burls, witches' brooms).

Galls and burls are similar in appearance while burls and witches' brooms have bud proliferation in common. Durst (1955) called bud proliferation "burl formation caused, by among other things, latent buds." Schmelzer (1977) suggested the cause of this bud proliferation to be mycoplasma-like organisms.

In contrast to Butin (1996, p. 225), who refers to a true gall when "…the bark is smooth and without excessive bud formation" and a burl when "the swelling encircles the entire stem," here galls and burls are conceptually separated because they are subject to different development processes and differ anatomically from each other. This conclusion is supported by Wagenführ (1989) who clearly distinguishes between ulcers and burls.

Tables 6.3 and 6.4 list several causes of these growth anomalies. There remains to be a conclusive explanation for certain symptoms of infestation (König 1957).

6.1.2.3 Prevention

Galls, burls, and witches' brooms are difficult to avoid. Either too little is known about their causes to take specific preventative action or it would require a disproportionate amount of effort to prevent the smallest wound. Witches' brooms in a tree's branches have little influence on the quality of its stem wood, making them irrelevant from the point of view of the wood processor.

Trees with malignant tumors should be removed as part of stand maintenance. They present entry points for further infection. Added value is no longer expected.

Table 6.3 Cause of burls and witches' brooms (selection)

Species	Responsible microbes and indirect influences
Gall (typically benign tumor)	
Oak (*Quercus* ssp.)	Oak gall midge (*Cecidomyiidae*) (Schwerdtfeger 1981) oak aphid (*Lachnus roburis*) (Altenkirch et al. 2002)
Spruce (*Picea abies*), pine (*Pinus sylvestris*)	Cause unknown (Schwerdtfeger 1981)
Ash (*Fraxinus excelsior*), birch (*Betula pendula*), maple (*Acer* ssp.), hornbeam (*Carpinus betulus*), fruit trees	Nectria fungi (Durst 1955; König 1957)
Red alder (*Alnus glutinosa*)	"Mono-bud-burl," parasitic fungi (Durst 1955); see also (Schwerdtfeger 1981)
Malignant tumor	
Spruce (*Picea abies*)	Nectria cancer *Nectria fuckeliana* (Altenkirch et al. 2002)
Larch (*Larix decidua*)	Larch cancer *Lachnellula willkommii* (Altenkirch et al. 2002)
Fir (*Abies alba*)	Fir cancer ("wheels-fir") *Melampsorella caryophyllacearum* (Altenkirch et al. 2002)
Hardwoods	Hardwood cancer (ash (Fraxinus) cancer, birch (Betula) cancer) *Nectria galligena* (Altenkirch et al. 2002)
Beech (*Fagus sylvatica*)	Beech cancer *Nectria ditissima, Nectria coccinea* (Altenkirch et al. 2002)
Ash (*Fraxinus excelsior*), poplar (*Populus* ssp.)	Bacterial ask cancer; *Xanthomonas* genus, *Pseudomonas* (Butin 1996)
Poplar (*Populus* ssp.)	Poplar cancer *Nectria galligena* (Butin 1996), *N. coccinea* (Durst 1955)
Oak (*Quercus* ssp.), red oak (*Quercus rubra*), beech (*Fagus sylvatica*)	Bleeding broadleaf crust *Stereum rugosum* (Butin 1996)

Table 6.4 Causes of galls, malignant tumors (selection)

Infested species	Responsible microbes and indirect influences
Burls	
Oak (*Quercus* ssp.)	Unknown (Butin 1996)
Black poplar (*Populus nigra*)	Bud proliferating mycoplasma-like organisms (Schmelzer 1977)
Sugar maple (*Acer saccharum*)	Parasitic fungi (*König* 1957)
Larch (*Larix decidua*), yew (*Taxus baccata*)	Unknown
Thuya (Thuja occidentalis)	Fire injuries are a precondition for thuya burls (Butin 1996)
Elm (*Ulmus glabra*), birch (*Betula pendula*), red alder (*Alnus glutinosa*), linden (*Tilia* ssp.), walnut (*Juglans regia*); also ash (*Fraxinus excelsior*), pear (*Pirus communis*), maple (*Acer* ssp.), oak (*Quercus* ssp.), cherry (*Prunus avium*)	Wagenführ and Scheiber (1989, p. 121) supposed the causes to be "tree species and breed related predispositions, injuries, insect infestations and other external influences"
Willow (*Salix* ssp.), poplar (*Populus* ssp.)	Tree tumor (*Agrobacterium tumefaciens*) (Butin et al. 2010)
Witches' brooms	
Common pine, North American pines (*Pinus* ssp.), juniper (*Juniperus communis*), cypress (*Cupressus* ssp.)	*Phytoptus* species (Gerard and Liernur 1927) on certain host plants, certain mistletoe species (*Viscum* sp.) trigger growth of witches' broom
Fir (*Abies alba*)	Fir broom rust *Melampsorella caryophyllacearum* (Altenkirch et al. 2002)
Spruce (*Picea abies*)	*Arceuthobium pusillum* (Gerard and Liernur 1927)
Larch (*Larix decidua*)	Rickettsien – similar bacteria (Nienhaus 1979)
Red alder (*Alnus glutinosa*), aspen (*Populus tremula*), elm (*Ulmus*), ash (*Fraxinus excelsior*), and beech (*Fagus sylvatica*)	Mycoplasma (Altenkirch et al. 2002)
Hornbeam (*Carpinus betulus*)	Ascomycete *Taphrina carpini* (Butin 1996)
Cherry (*Prunus avium*)	Ascomycete *Taphrina cerasi* (Butin 1996)
Birch (*Betula pendula*)	Ascomycete *Taphrina betulina* (Butin 1996)

Galls and burls are a different matter. Quality burl wood is highly valued for its attractive figure. So much so that birch (Betula) burl wood has successfully been produced artificially by cross and vegetative breeding (König 1957). No such success has occurred using other tree species, e.g. walnut (Juglans regia) (Gottwald 1983).

6.1.2.4 Impact on Use

Galls and *burls* are undesirable if their small size, irregular surface, enclosed bark, or rotted areas prohibit the wood from being used for a specific purpose. This includes all malignant tumors.

Although burls occur in many species, only high-quality burl wood is prized for its unique color, structure, and texture. Conditions for high-quality burl wood are:

- Contained, compact exterior form
- No ingrown bark or mineral galls
- No rotting
- No major branch penetration
- A minimum size that enables the wood to be used for a specific purpose

Fig. 6.12 Veneer with latent buds, birds eye maple (*Acer saccharum*)

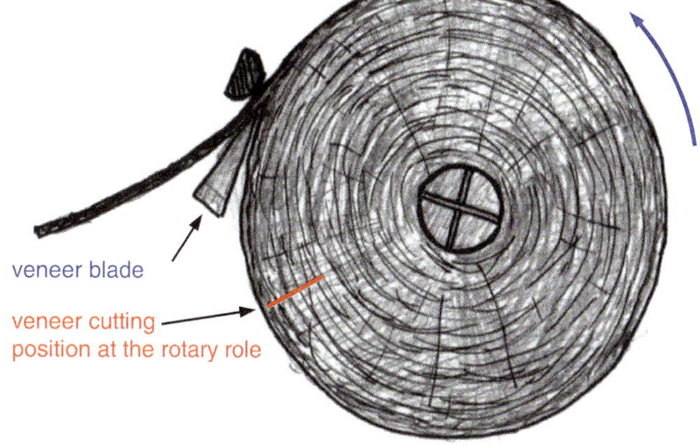

Fig. 6.14 Centrical peeling of the bird's eye maple (*Acer saccharum*)

Fig. 6.13 Veneer section (radial) from a walnut (*Juglans regia*) burl

The most important indigenous hardwoods with attractive burl are maple *(Acer saccharum)* (bird's eye maple, Fig. 6.12); ash (*Fraxinus* ssp.*)*, especially olive ash (*Fraxinus*); poplar (*Populus* ssp.*)* (mappa); oak (*Quercus* ssp.); and elm (*Ulmus glabra*). In pines (*Pinus sylvestris*), yew, red cedar, and redwood burl (*Taxus baccata, Thuja* ssp., *Sequoia sempervirens, Sequoiadendron giganteum*) (vavona) are most valued. Prized among (sub) tropical woods are madrone *(Arbutus menziesii*), makamong (*Afzelia xylocarpa*), myrtle (*Umbellularia californica*), walnut root burls (*Juglans regia*) (Fig. 6.13), and padouk (*Pterocarpus* ssp.*)* (Wagenführ 2007). Small galls from the tree heather (*Erica arborea*) are used to make the famous Bruyére pipes. Birch (*Betula* ssp.*)* burls are known as "semi masur," because they lack the decorative effect of the latent "eyes" (Matzek 1983).

It is important not to be overly optimistic about the possible proceeds earned from selling burl wood (Schmieder 2003). In German forests, "normal" galls and burls are rarely so perfect that they yield top price from a veneer manufacturer, as may be the case, for example, with a special walnut (*Juglans regia*) burl on the international timber market. "The tale of the high wood prices" (Griese 2004) shows that the hidden risks in burl wood keep it from yielding above average prices.

Witches' brooms are economically insignificant. Their cancerous growths occur usually in the tree branches.

6.1.2.5 Technological Adaptation

Stem burls as on bird's eye maple *(Acer saccharum*), elm (*Ulmus* ssp.), ash (*Fraxinus excelsior),* poplar (*Populus* ssp.*)*, and to a certain extent also walnut root burls (*Juglans regia*) are centrically peeled, in order to optimally reveal the decorative effect of the figure (Fig. 6.14).

Basal burls are typically divided into segments and sliced or peeled (eccentric). The quality of the burl only becomes apparent during processing. The segment breakdown requires a great deal of experience in order to take full advantage of the burl wood and maximize yield (Furniermagazin 2004; Holztechnik Fachkunde 2003, p. 119–122).

The irregular fiber orientation can cause warping and fissures in the drying veneer. To avoid cracks and breaks, the fresh veneer is dried very gently (ironed). The tendency to warp during storage increases with higher moisture levels.

6.2 Impact of Animals/Humans

6.2.1 Vertebrates: Browsing, Fraying, and Stripping Damage (Richter and Prien 2006)

6.2.1.1 Description

As long as humans have been managing forests in an effort to maximize timber yields and improve wood quality, foresters and hunters have shared a conflict of interest, the former prioritizing timber harvest and the latter wild game stocks. Forest owners are not categorically against all types of game that could affect tree growth and wood quality, but many do oppose an overpopulation of deer and other browsing species. Hunters, for their part, understand that excessive deer populations are harmful to trees and realize that a forest unable to flourish jeopardizes healthy game stocks, thereby limiting the number of kills and the chances of bringing home a prized hunter's trophy.

Yet, what is an "excessive game population"? Research has shown that a realistic estimate of the game stocks lies in the subjective "want" or "not want" of the hunter (Prien 1997). The conflict of interest is compounded when the hunter is not the forest owner or when the game population in a forest reaches levels beyond the forest owner's control.

Not that the existence of wild game animals in the forest should be questioned – they are of course an integral part of forest biocenosis. However, to establish a healthy balance between the interests of foresters and hunters, one must rely on facts. Browsing, fraying, and stripping damage needs to be weighed against the profits generated from hunting (including trophy hunting) in the context of necessary game population management and, not least, as a "recreational hobby."

From the perspective of the woodworker, all damage is of importance if it affects the quality of wood at harvest:

Browsing damage is inflicted when roe deer (*Capreolus capreolus*), red deer (*Cervus elaphus*), fallow deer (*Dama dama*), mouflon (*Ovis orientalis musimon*), and rabbit (*Lepus europaeus*) nip off the buds and shoots of young trees. It can lead to stem deformations and forked growth (Fig. 6.15). Less common is bud and shoot browsing by squirrels (*Sciurus vulgaris*) and certain birds such as crossbills (*Loxia pytyopsittacus*).

See Plate 6.7 for browsing damage photos.

Fraying and rubbing damage is caused by deer (*Cervidae*) rubbing their antlers against the trees, resulting in bark, stem, and branch injuries which can lead to stem deformations (Fig. 6.16). Wild boars (*Sus scrofa*) can also inflict bark injuries and stem deformations on the so-called rubbed trees.

See Plate 6.8 for fraying and rubbing damage photos.

Stripping damage refers to injuries such stem deformations, resin glut, decay, and death of trees caused by animals (*Mammalia*) using their teeth to strip the bark off young smooth-barked trees (Fig. 6.17).

See Plate 6.9 for bark stripping damage (deer) photos.

These types of damages are also caused by animals living in forests of the subtropics and tropics.

Fig. 6.15 Pine (*Pinus sylvestris*) with old browsing or rubbing damage. Side shoots take on a lyre shape

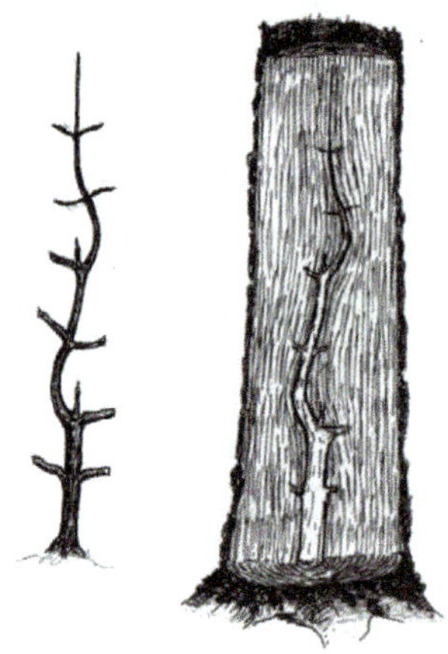

Fig. 6.16 Fraying damage in a pine (*Pinus sylvestris*) (*left*) and its location in the mature stem (*right*)

Fig. 6.17 Summer time bark stripping on a Douglas fir (*Pseudotsuga menziesii*)

6.2.1.2 Causes

Commercial forests should strive for an ecologically and economically sustainable wildlife density. Meaning a wildlife density "… by which the natural and artificial regeneration and growth of the predominant tree species or prevailing agricultural crop in the respective hunting area takes place with as few injuries is possible. At the same time, it is assumed that silvicultural and hunting activities meet the minimal requirements with regard to browsing improvements and supplemental winter feeding" (Prien 1997,

Spruce beech (*Picea-Fagus*) regeneration fully browsed over by deer (*Cervus elaphus*). Game density:14 animals/100 ha forest floor (Katzhütte, Germany)

Fir (*Abies alba*) with roe deer (*Capreolus capreolus*) "bonsai" browse

Mountain ash (*Sorbus aucuparia*) with roe deer (*Capreolus capreolus*) browse

Maple (*Acer pseudoplatanus*) damaged by a rabbit (*Lepus europaeus*)

Oak (*Quercus* ssp.), birch (*Betula pendula*), and pine (*Pinus sylvestris*), damaged by a beaver (*Castor fiber*) (Königsbrück, Germany)

Aspen (*Populus tremula*) damaged by a beaver (*Castor fiber*) (Mo i Rana, Norway)

Fir (*Abies alba*) with squirrel (*Sciurus vulgaris*) bud browse

Tapir (*Tapirus bairdii*) browse on a year old tropical sapling (Surinam)

Plate 6.7 Browsing damage

Dead young Douglas fir (*Pseudotsuga menziesii*), with deer (*Capreolus capreolus*) fraying damage (Germany)

Dead Douglas fir (*Pseudotsuga*), with red deer (*Cervus elaphus*), rubbing damage (Germany)

Larch (*Larix decidua*) with deer (*Cervus elaphus*) rubbing damage (Germany)

Reindeer (*Rangifer tarandus*) butts a young pine (*Pinus sylvestris*) (Lappland, Finnland)

Larch (Larix decidua) with moose (*Alces alces*) rubbing damage (Darlana, Sweden)

Deformed rubbing tree (*Pinus sylvestris*) near a wild pig (*Sus scrofa*) wallow (Tharandt, Germany)

Total loss caused by wild pigs (*Sus scrofa*) in a spruce stand (*Picea abies*) (Tharandt, Germany)

Basal stem injuries to a group of alder (*Alnus glutinosa*) by wild pigs (*Sus scrofa*) (Tharandt, Germany)

Grizzly bear (*Ursus arctos horribilis*) paw damage on a *Pinus contorta* (Yellow Stone, USA)

Plate 6.8 Fraying and rubbing damage

Summertime bark stripping on a 20-year-old spruce (*Picea abies*) by a deer (*Cervus elaphus*)

Summertime bark stripping on a Douglas fir (*Pseudotsuga menziesii*) by a deer (*Cervus elaphus*)

Summertime bark stripping on a beech (*Fagus sylvatica*) by a deer (*Cervus elaphus*) (Photo: S. Prien)

Winter bark stripping on spruce (*Picea abies) by a deer* (*Cervus elaphus*) (Katzhütte, Germany)

Winter bark stripping on mountain ash (*Sorbus aucuparia*) by a deer (*Cervus elaphus*), (Katzhütte, Germany)

Left: Old, unsealed stripping wound on a 60-year-old spruce (*Picea abies*) with red rot (*Heterobasidium annosum*), caused by a deer (*Cervus elaphus.*) (Thuringia, Germany). *Right*: 80-year-old, fully stripped stand (Photo: S. Prien).

Bulging calluses caused by a deer (*Cervus elaphus*) stripping on a 100-year-old spruce (*Picea*) (Thuringia, Germany)

Spruce (*Picea abies*) stem broken at a spot damaged by deer stripping (*Cervus elaphus*) (Erz Mountains, Germany)

Plate 6.9 Bark stripping damage (deer)

Table 6.5 Economic viable wildlife density per 100 ha of forest land based on the main wild game species and feeding conditions (Müller 1963)

Feeding conditions Main species[a]	Very poor	Average[b]	Very good
Deer (*Cervus elaphus*)	1.0	1.4	2.0
Fallow deer (*Dama dama*)	–	3.0	8.0
Mouflon (*Ovis orientalis musimon*)	3.0	5.0	8.0
Roe deer (*Capreolus capreolus*)	4.0	6.0	9.0
Example[c]	4.5	8.4	14.5

[a]If the species is a subspecies, then the wildlife density should be divided in half
[b]Medium-sized feeding conditions, 300–500 m altitude, >60 % spruce (*Picea abies*), <assuming 35 % eatable shrub layer
[c]Upper limit of wildlife density, if deer (*Cervus elaphus*) is the main species and other species are subordinate[a]

Fig. 6.18 The rubbing damage to this 14-year-old larch (*Larix decidua*) remains unsealed after 70 years

p. 137). In this context, Prien defined "tolerable" as grazing that affects up to 10 % of the plants of a cultivated area and stripping damage affecting 10–15 % of the prevailing and dominant trees (classes I and II of the main tree species according to Kraft (1884)). The game densities listed in Table 6.5 are based on these limits and depend on the type of game and the feeding conditions.

Attention! If these limits are exceeded for hunting purposes, either intentionally or due to an absence of hunting, then the balance between food supply and consumption is upset and the result is an intolerable level of damaged trees.

In the old growth forests of the boreal and temperate regions, but also in the primary rain forests of the tropics, there is usually a natural balance between food for herbivores and their "hunters." This is not the case in the savanna and subtropics. There, periods of drought can upset the balance between food supply and food demand by herbivores. Damage to trees and saplings is often the result.

Commercial forests in temperate zones are susceptible to the following types of damage:

Browsing Damage: Especially roe deer (*Capreolus capreolus*), hare (*Lepus europaeus*) and wild rabbits (*Oryctolagus cuniculus*), sometimes deer (*Cervus elaphus*) and mouflons (*Ovis orientalis musimon*), and less frequently fallow deer (*Dama dama*) nip off the little buds and terminal shoots of young trees between 0.3 and 1.5 m in height because for their rich nutrients. Particularly in trees with opposing branch formation, the loss of the terminal shoot can lead to permanent forking. In alternate sprouting tree species and softwoods, the shoot deformation usually will coalesce.

Fraying and Rubbing Damage: Male red and roe deer (*Cervus elaphus, Capreolus capreolus*) free their antlers from velvet, by rubbing against young, flexible trees between 2 and 10 cm thick. Bucks also rub against the trees during the rutting season as a form of aggression and to mark their territory (Fig. 6.18). The damage usually occurs at stem heights between 0.5 and 2 m. Fraying and rubbing damage is most common on softwoods and rare on immature and free-standing trees. If the trees survive, they typically exhibit stem curvature, callusing, and sometimes decay.

Wild pigs (*Sus scrofa*) will also rub up against the so-called "rubbing" trees after wallowing, and boars will rub the trees to mark their territory. The bark is often completely damaged around the stem and can lead to the localized demise of entire tree stands (Briedermann 1986).

Bark Stripping: Red deer, sika deer, elk, mouflon, and (sometimes!) also fallow deer (*Cervus elaphus, Cervus nippon, Alces alces, Ovis orientalis musimon, Dama dama*) have a species and population-related tendency to strip off bark with the incisors of their lower jaw for nourishment. Bark stripping usually occurs at stem heights of up to 2 m. All tree species are susceptible to stripping, while some are more preferred (Prien 1997). Stripping damage is particularly heavy in large-scale single species stands, in the winter when the nutrient supply is low or when the animals are frequently disturbed or stressed.

The wounds become infection portals for decay. Stem interiors and exteriors become weak. The trees try to seal off or grow beyond the damage. This leads to misshaped stem calluses with deep rot penetration (red rot (*Heterobasidion annosum*) in spruce (*Picea abies*), white rot in beech (*Fagus sylvatica*) (Agaricomycetes) and valuable hardwoods). Severely damaged trees are especially vulnerable to snow and wind (Fig. 6.19).

6.2.1.3 Prevention

To a certain extent, the management of wild ruminant animal populations remains unpredictable and random. Therefore, damage cannot fully be avoided. It can, however, be kept to low, economically, and environmentally tolerable levels. This requires a scientifically sound strategy and relentless consistency in practical implementation – free from wishful thinking and lobbying.

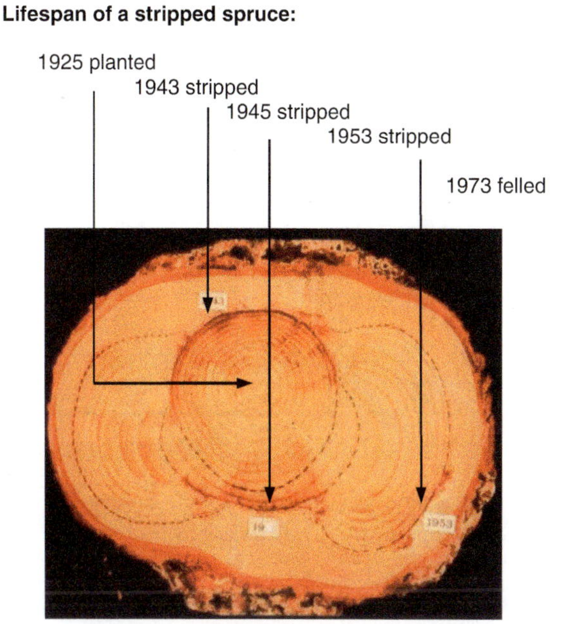

Fig. 6.19 Spruce (*Picea abies*) with callused stripping injury and onset of rot caused by deer

Damage from wild game can be reduced using up-to-date, future-oriented, and complex management and protection measures (Prien 1997). Since any economically significant damage to a forest by wild game typically has multiple causes, damage control generally requires taking both preventative and protective measures.

Sustained wild life management by quantity (kill quotas, game species) and quality (gender ratio, age group) is the most important prerequisite for the effectiveness of other damage-reducing measures, such as improving the natural food supply and providing supplemental winter feedings (Prien 2006). A good reference measurement for sustainable damage-reducing game regulations is the tolerance limits for browsing and stripping established as part of the forest management appraisals of the respective administrative unit (forestry, forest farm, private forest enterprise, etc.).

An intensively cultivated, often densely planted, and thus typically heavily shaded commercial forest often has a low natural nutrient supply, depending on the site and age of the stand. Here browse and stripping damage is managed by creating permanent feeders and cultivating feeding plots and by providing woody browse cuttings, mast trees, and supplemental winter feeding in the mountains. However, habitat improvements and supplemental feeding will in turn lead to an increased game population that will ultimately need to be reduced.

In many places the wildlife is under such stress, it significantly disturbs their natural biorhythms, particularly the species-specific rhythm of nutrient intake and rumination. The stress caused by having to suddenly flee for protection leads to a sharp rise in energy expenditure and higher metabolism rates. As a result, the amount of browsing and stripping damage increases. Measures that reduce disturbances and stress are an integral part of damage control: Rules of behavior (public relations) need to be enforce, game preserves established, hunting regulated (an appropriate level of sport hunting), and reviver supervision increased.

Because wild animals have very different food preferences for individual tree and shrub species, the type-specific behavior of wildlife in their choice of food can be exploited by offering more preferred woody browse in the stands and especially in the vicinity of the feeding plots. Doing so distracts the wildlife away from browsing and stripping the more vulnerable commercial trees.

Several diversion measures have proven particularly effective. Bushes and woody browse can be deliberately planted in specific sites, for example, depending on the location, willows (*Salix* ssp.), hornbeams (*Carpinus betulus*), oaks (*Quercus* ssp.), fir (*Abies alba*), and elder (*Sambucus* ssp.). Supplemental woody browse cuttings can be provided, such as mountain ash (*Sorbus aucuparia*), sallow (*Salix caprea*), aspen (*Populus tremula*), and deciduous species. And trees can be felled then left in the forest as a nutrition source, in areas with heavy browsing or stripping where there are no cultivated feeding plots or insufficient supplemental woody browse. Material removed in thinning is particularly suitable as supplemental feed, for example, poplar (*Populus* ssp.), mountain ash (*Sorbus aucuparia*), common ash (*Fraxinus excelsior*), pine (*Pinus sylvestris*) and spruce (*Picea abies*). The latter measures are particularly effective for reducing winter bark stripping by deer (*Cervus elaphus*) because, in comparison to the standing trees, the felled trees are five times more susceptible to bark stripping.

Despite consistent wildlife management, improved nutritional supply, and lower habitat disturbance, a degree of risk always remains. Commercially unacceptable damage will continue to occur as long as wild game animals continue to browse and strip bark in their search for nourishment and inflict injuries through other types of natural behavior such fraying and rubbing. Simply put, wildlife is wild. Therefore, certain forest sites and tree species will also continue to require the use of special protective measures:

- Browse and fraying damage: Protection by various types of guards (Fig. 6.20), chemical repellents, and fencing (Fig. 6.21).
- Stripping protection: Mechanical processes (green and dry cover, polyethylene nets, silicate sand paint, Fig. 6.22), mechanical-biological processes (artificial bark injury – wound cork formation/hardening, Fig. 6.23), and chemical procedure (application of repelling smell and tastes on of selected trees).

It is important to accurately assess which type of wild animals is causing the browsing or stripping damage in a plantation, natural regeneration, or immature stand in order to maintain the cost of expensive protective measures.

In this context, it is important to clearly establish who is responsible for paying for the extra expense of the additional protective measures. In the case of agricultural field damage, the game tenants are responsible. The causative principle should also apply in the forest.

Fig. 6.20 Special trunk guard material

Fig. 6.21 Protective deer fencing

Fig. 6.22 Bark stripping protection: Latex and sand coating (*left*). Right stripped beech (*Fagus sylvatica*)

6.2.1.4 Impact on Use

Browsing damage impedes growth in height and leads to species segregation in natural regenerated stands. The seedling stage lasts longer, the targeted diameter is reached later, and hence, the stand's turnover time is longer. Severe branchiness in the lower section of the stems and abnormal structures in the juvenile wood prevent the wood from being used for high-value applications.

Fig. 6.23 Mechanical, biological stripping protection: resinous spruce (*Picea abies*) stem done with a bark scraper tool

Fig. 6.24 A 15-year-old stripping injury disrupts the structure of the wood of a Douglas fir (*Pseudotsuga menziesii*)

Fraying and rubbing damage is typically callused over but can also cause forking, stem rot, and disfiguring in the wood structure in the lower stem area (see Figs. 6.15 and 6.16).

If trees manage to survive the rubbing injuries caused by wild pigs (*Sus scrofa*), they are typically dirty at the base, resinous, and often so badly damaged that there use is limited.

Bark stripping damage leads to decay which in turn often causes stem to break. Associated usages cause loss of quality (decay, fracture zones in the wood, disturbed wood structure, resin flow). Callused injuries lead to ingrown bark and deformations on the stem (Figs. 6.24 and 6.25).

An average time period of about 10 years can be used as a benchmark for determining when decay began in a spruce (*Picea abies*) – depending on the degree of stripping and the site conditions. After the protective barrier of the living sapwood has been penetrated, the decay will progress in the mature wood between 10 and 20 cm each year.

In a 120-year-old spruce injured 100 years ago by bark stripping at a height of 1 m, an annual decay of 10 cm would result in an area of decay extending 10 m up the stem and 1 m down. The wood quality grades would be distributed as shown in Fig. 6.26 (HKS 2002a).

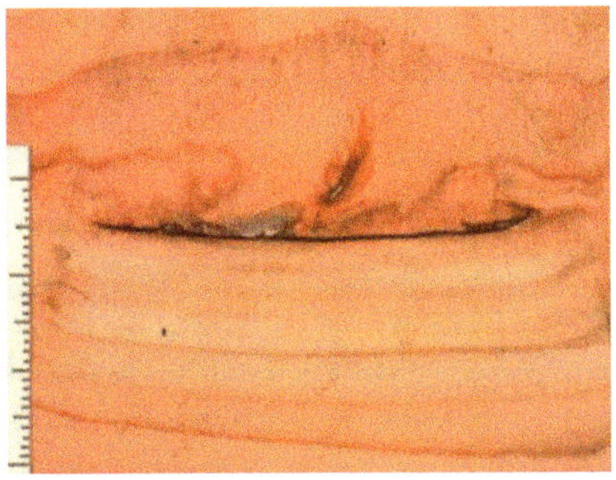

Fig. 6.25 Larch *(Larix decidua)* boar with ingrown bark as a result of stripping damage

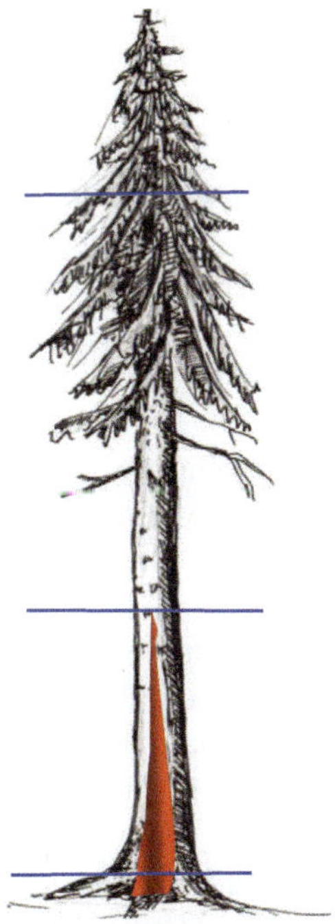

Fig. 6.26 Grading scheme in a spruce (*Picea abies*) with red rot

The financial impact of long-term insufficient wildlife management and the accompanying damage to timber yields was studied on a 70-year-old, fully stripped spruce stand in a state forest in Thuringia, Germany (Autorenkollektiv 2006). The sustained deer density in this case was 12 animals/100 ha. That corresponds to approximately ten times more than the economically viable deer (*Cervus elaphus)* density of a mid-sized hunting area (see Table 6.5). Alone the sorting of the lower stem section B to less valuable D quality result in a financial loss of 50 €/ha and year. Revenue from game and timber leases were offset by 8 €/ha and year. The difference corresponds to a loss to the forest owner or a tribute to an antiquated game keeping and hunting ideology.

Binder (2008) has developed appropriate tables for the collection and evaluation of bark stripping damage to spruce (*Picea abies*).

Fig. 6.27 Felling damage to three logs

6.2.1.5 Technological Adaptation

Sorting grade sections of round wood from normal B/C quality in D grade takes considerable time and effort. Pruning unhealthy sections from trees results in shorter logs (ISK) and a significant amount of slash. If a section of the D grade wood is kept, there is always the risk that parts of the wood will be rotten.

However, using special cutting equipment, heavily decayed pine wood from branch-free stem sections can still be processed for quality lumber. In the former GDR knotless spruce (*Picea abies*) was harvested as D2 grade timber. It could have red rot centered in 50 % of its mid-diameter or an off-centered crescent-shaped rot area (TGL 1988).

These days, decay can be detected in standing trees using noninvasive diagnostic instruments (sonic tomography (Weiß 2006), magnetic resonance imaging). Currently, progress is being made on the large-scale application of such instruments. Problems in implementation occur in distinguishing and differentiating between defects accurately and quickly enough to take appropriate measures.

The extra effort needed for sorting standing trees continues on in the manufactured wood (disruptions in the course of fiber, ingrown bark, decay, calluses) (DIN 2001a, DIN 2002b, DIN 2004a).

The production of chipboard, fiberboard, and paper requires limits in the amount of D grade wood added.

6.2.2 Forestry Operations: Felling and Hauling Injuries (Plant Exudates)

6.2.2.1 Description

Felling damage refers to mechanical injuries caused by humans during timber harvest to the branches, bark, and wood of standing trees, with long-term impact on wood quality as well as to the felled logs (breakage, split stems, wood fissures) (Fig. 6.27).

See Plate 6.10 for felling injuries photos.

Fig. 6.28 Hardwood with an old, unhealed hauling injury

Hauling damage refers to mechanical injuries caused by humans during timber harvest to the bark and wood in the lower (most valuable) bole section of standing trees by logging equipment as well as to logs being hauled out of the forest with long-term impact on quality (Fig. 6.28).

See Plate 6.11 for hauling photos and Plate 6.12 for stacking photos.

Commercial forests require repeated maintenance interventions to remove poor-quality trees and promote trees in optimal condition (stem and crown care). Yet, when improperly conducted, these interventions can cause significant stem damage to standing trees and felled logs. Over the long term, felling and hauling injuries can cause even greater harm to the quality of a forest stand than natural biotic influences (rot, insect infestation, stripping and rubbing damage caused by animals).

Despite "best practices," felling and hauling injuries in temperate forests can significantly affect the quality of the remaining trees in a stand and the harvested timber. General factors included:

1. Maintenance frequency: From a forestry yield science perspective, relatively frequent, but low-level maintenance interventions are most beneficial. Frequent maintenance, however, also increases the risk of injuries to the trees remaining in the stand. From a forestry operations perspective, the best approach is as few interventions as possible with large harvests (piece-volume-law). The risk of stem injury is reduced. However, growth loss is still likely due to the less than optimal maintenance of the trees' basal area.
2. Principle of selective harvesting: Selective harvesting (removal of individual trees with sufficient diameters) is increasingly becoming standard practice in Central Europe, replacing other timber harvest practices (clear cutting and compartment shelter wood systems).

   However, selective harvesting increases the risk of hauling injuries to the remaining trees.
3. Stand structure: As the mix of tree species, vertical density, and number of individual trees in a stand increase, so does the risk of damage to standing trees during forest thinning operations.
4. Timber harvesting: For the most part, the days of felling timber by hand and hauling logs away by horse have passed; today trees are harvested using powerful logging equipment, i.e., harvesters, hauling winches, and forwarders. And yet, each logging operation must be tailored to the specific site conditions of the individual stand.

The above factors have a particularly significant impact in primary and secondary forests of the tropics with their high stand density and great variety of tree species. For example, an estimated 800 different tree species grow in the primary forests of Surinam. Only about 350 have had their anatomical, physical, and technical properties documented. Around 30 of these are actually used, of which there are five main types of wood (Comvalius 2010). There is currently no in-depth research available on accepted levels of intervention and maintenance that does not effect biodiversity, Therefore, based on the ten principles for certification of forestry companies established by the Forest Stewardship Council (FSC) (N N 2009b) and the "code of practice," (Code of Practice 2011) a logging limit was set of 25 m^3 stem wood (crown components are not included) every 25 years for a defined, undeveloped forest area. The minimum distance between two logged trees must not be less than 10 m. The coordinates of each stem are determined using GPS. This makes it possible to track the wood to the end user. Trees on the "red list" may not be logged. Transport roads and skid trails are permissible (Richter 2012).

Plant exudates refer to wound sealants on woody plants. Certain pine species heal their injuries with a resinous exudation consisting of 22 % turpentine oil and 70 % rosin. This resin is harvested worldwide as a forest byproduct and widely used in the chemical and pharmaceutical industries. In former East Germany and the Eastern Bloc, the limited availability of other more resinous tree species led to the large-scale resin tapping of the local pines (*Pinus sylvestris*). These pines, however, yielded only 25 % of the resin available from tropical pine species in Central America and the Caribbean (*Pinus elliottii*) (Stephan 1973) even despite the use of resin flow stimulants (sulfite lye, yeast extract).

Resin tapping involves cutting one or two grooves in maximum two thirds of the stem circumference. A third remains as a living strip. In the vegetation period, every 4–6 days, a fresh notch about 3 mm deep is made in the groove. Both the horizontal and vertical resin canals are cut and the resin flows out into a receptacle (Fig. 6.29). The resin yield can amount to 100–150 g per resin tap. After a maximum of 12 years of resin extraction, a pine stand is completely tapped and ultimately logged (Stephan 1973; Stephan 2012).

Felling a tree without a proper face notch is unprofessional, is risky, and results in a lower wood volume and quality (Saxony, Germany)

The forked birch (*Betula pendula*) stem fell in the wrong direction permanently damaging the neighboring spruce (*Picea abies*)

Beech (*Fagus sylvatica*) with fresh bark injuries during a winter felling

Beech (*Fagus sylvatica*) with successfully callused felling injury

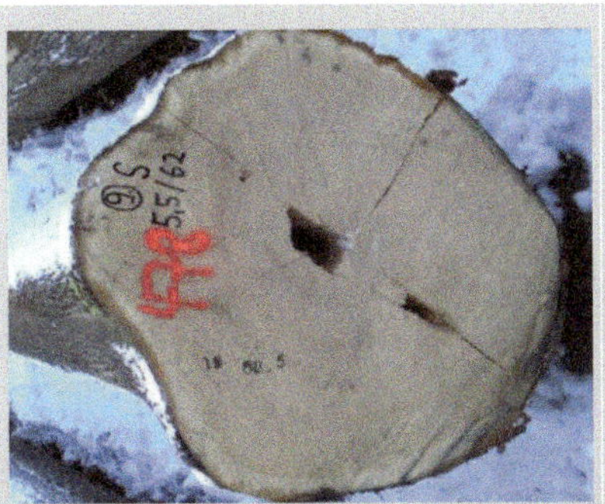

Too flat of a face notch left this stem hanging. A 50 cm long wood splint was taken from this oak (*Quercus rubra*) (mass adjustment required!)

The buttress roots of this valuable greenhart stem (*Tabebuia serratifolia*) were cut off. Left felling injury, *right* back hauling damage to stems (Surinam)

Two closely neighboring kopi stems (*Goupia glabra*) were felled against FSC guidelines causing serious damage (Surinam)

Buttress roots are left standing which could have prevented future cracks in the logged wood (Surinam)

Growth cracks in a donsedre (*Cedrelinga cateniformis*) log appear because the back cut was not made at the root base and the log left unfastened (Surinam)

Plate 6.10 Felling injuries

This harvester injured a spruce (*Picea*) stem (insufficient spacing, no skid trail) (Hessen, Germany)

Basal stem damage to this spruce (*Picea abies*) resulted in decades of decay (*Heterobasidion annosum*) followed by ant infestation (*Formica* ssp.)

Unsuccessful attempt to callus over a hauling injury, combined with resin bleeding on an old pine (*Pinus sylvestris*)

Larch (*Larix decidua*) injury inflicted in the pole wood stage has successfully callused. Ingrown bark and a defective wood structure have remained

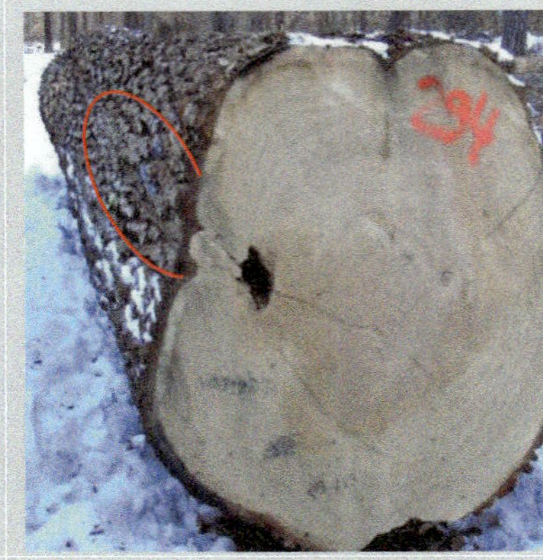

An old, callused over hauling injury depreciates one side of this veneer oak (*Quercus petraea*)

Beech (*Fagus*) with fresh hauling injuries to the stem base

Successfully callused hauling injury on a beech (*Fagus*) (compartmentalization and red heartwood formation)

Old hauling injury to the stem base of this beech (*Fagus*) led to D grade classification, higher risk of breakage

Hauling injury on an alder (*Alnus glutinosa*) with ingrown bark, discoloration, and decay

A kopi stem is being hauled by using a tree standing nearby as a pully guide. The stem is severely damaged and rot is likely (Surinam)

Plate 6.11 Hauling

Left: In temperate regions, the hauling damage done to this ash (*Fraxinus excelsior*) can still heal (Germany).
Right: This is not the case in the tropics, 12-year-old haulinging injury on a basralocus (*Dicorynia guianensis*) (Surinam)

Hauling and stacking logs in mud without a platform leads to future technical damages during processing (tool wear and tear). A poor practice example found in Spessart (*left*), Germany, as well as in the tropics (*right*) (Surinam)

Plate 6.11 (continued)

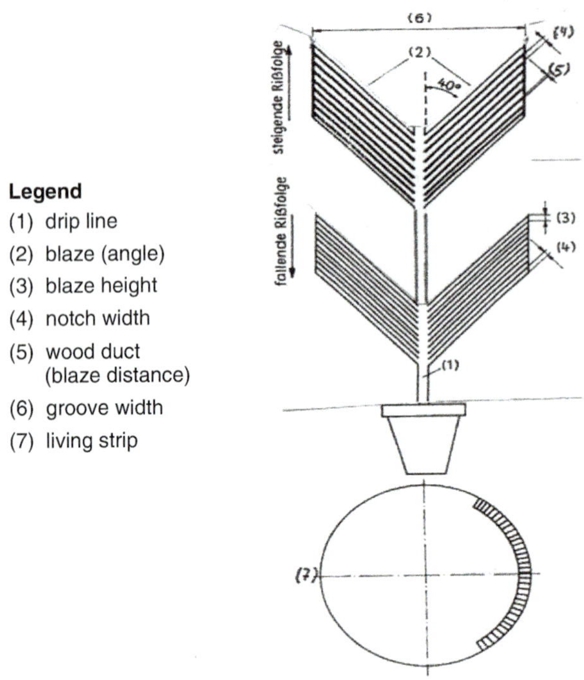

Fig. 6.29 Schematic diagram of a resin groove using the Chorin method (in reference to Stephan (1973)). *1* drip line, *2* blaze (angle), *3* blaze height, *4* notch width, *5* wood duct (blaze distance), *6* groove width, *7* living strip

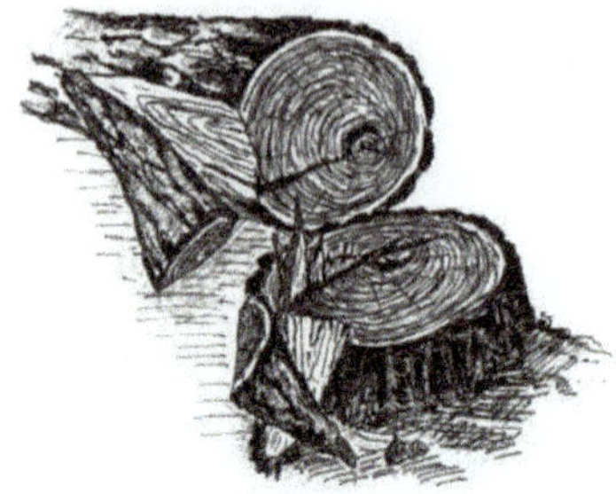

Fig. 6.30 Larch (Larix) stem with split fiber and cracked sapwood a result of a missing sapwood cut and a missing hinge

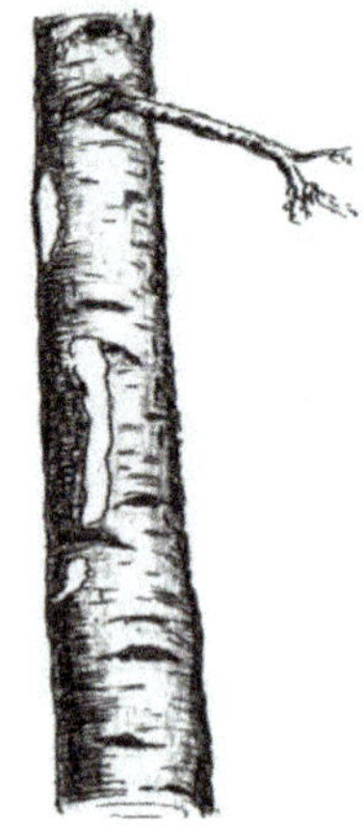

Fig. 6.31 Bark injury on a hardwood after summer felling

Fig. 6.32 Bark injury caused by a cable tied around the stem base

Other pine species used to for resin tapping are as follows: European black pine (*Pinus nigra*), Rumelian pine (*Pinus peuce*), and panzer pine (*Pinus leucodermis*). Resin from Aleppo pines (*Pinus halepensis*) has historically been used to make retsina wine.

Spruce (*Picea abies*) resin is extracted by scraping the pitch from bark injuries. The scraping often causes further injury to the tree resulting in decay.

Tropical trees produce wound sealants as water-insoluble resins, as water soluble or water-insoluble gums, tannin containing kinos, or polyterpene-based latices (Lange 1998a, b, c). Depending on the method used, the trees are either torn, tapped, or notched in order to extract the forest byproduct.

See Plate 6.13 for use of plant exudates (forest by product) photos.

6.2.2.2 Causes

Felling and Hauling Injuries: Damage occurs when maintenance operations are undertaken without previously conducting a thorough stand inventory. The absence of established skid trails and logging roads makes it difficult to determine in which direction the trees should be felled and trees end up being hauled through the middle of the remaining stand.

Often operators lack the skills and know-how required for timber harvesting (knowledge of the idiosyncrasies of the species, ability to judge weather conditions, skills to operate chain saws (Fig. 6.30) and other logging equipment).

If the forestry equipment used is not adapted to the specifics of the stand (distance between trees, stem diameters, type of terrain, etc.), injuries are unavoidable.

Trees are frequently harvested in the growing season. This is especially common when there is a year-round timber contract, but no arrangement for cold deck storage or chemical preservation (Mahler et al. 1997). During the growing season, smooth-barked trees are particularly vulnerable to bark injuries (Fig. 6.31).

In many cases, logging contracts do not sufficiently enforce sanctions for failure to comply with the quality requirements for logging, hauling, and storage along forest roads (Fig. 6.32).

Callused injuries on an oak stem (*Quercus* ssp.) near a log piling. The expected B quality is degraded to D quality	Spruce (*Picea abies*) used to support log piles exhibiting red rot (*Heterobasidion annosum*). Degraded from B/C to D quality	Damage to a larch log (*Larix decidua*) made by a grip forwarder during piling in Germany is similar to damage to a wana (*Ocotea rubra*) made during piling at a sawmill in Surinam	

Plate 6.12 Stacking

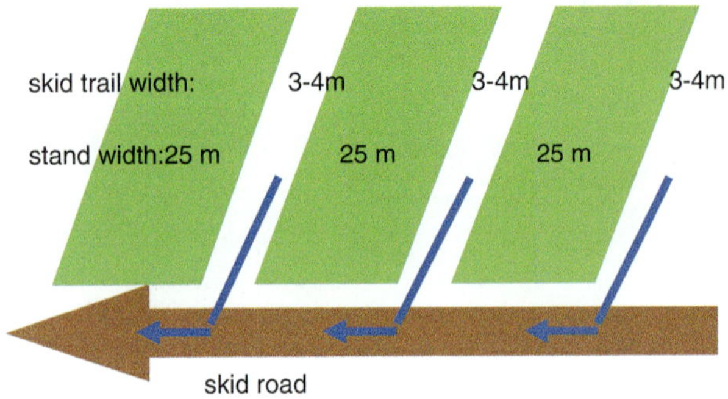

Fig. 6.33 Inventory of an immature softwood stand

To minimize effort, back cuts are sometimes incorrectly made through the (buttress) root collar. Felling in this manner leaves the log susceptible to stress cracks.

Logging in the primary rain forests of the tropics naturally differs from logging in temperate forests. In the rare case of selective single stem harvests, usually the strongest and highest-quality stems are logged without taking into account the balance of tree species in the area (N N 2009b). Closely neighboring tree crowns and the network of lianas are damaged in the process. The tree is typically topped under the first main branch. The usually long, branch-free stems of more than 20 m are difficult to move. Standing trees used as pully supports during hauling occur, lasting damage to their trunks.

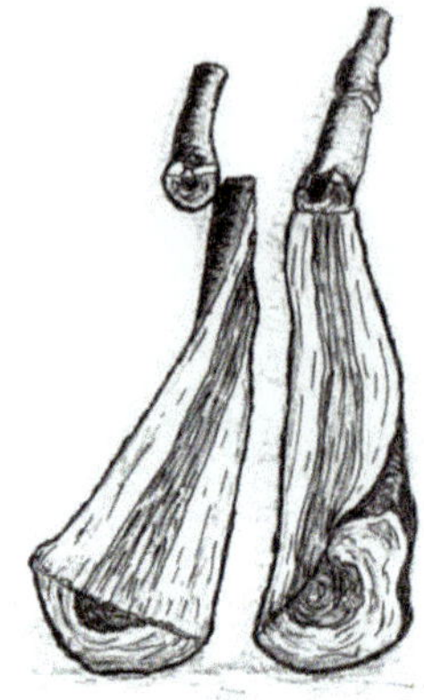

Fig. 6.34 Beech (*Fagus sylvatica*) stem wood degraded to industrial wood – incorrect felling method caused the fork to crack down the center

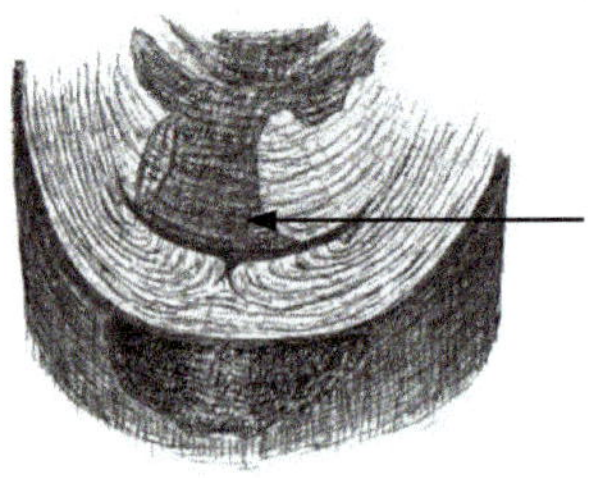

Fig. 6.35 Healed hauling damage on a beech (*Fagus sylvatica*)

6.2.2.3 Prevention

Felling and Hauling Damage: In commercial forests, it is important to take a stand inventory before the start of any logging operation to insure that the harvesting equipment can maneuver freely without permanently damaging the roots of unharvested trees (Fig. 6.33). Skid trails must be permanently marked, deck plots clearly designated, and buffer trees should be left standing to shield other trees from hauling and storage damage.

See Plate 6.14 for stand inventory photos.

Since harvesting damage is typically due to human error, particular emphasis should be placed on training. Felling and hauling work should only be conducted by well-trained personnel (Fig. 6.34).

The felling and hauling methods used must be tailored to the size of the trees, stand density, and the soil conditions. In high density stands (tropical primary forests!), logs should be bucked on site to prevent hauling damage.

Summer harvesting in temperate forests should be avoided when possible because even the slightest stem contact could lead to bark injuries.

Timber harvest contracts need to include sanctions for quality loss.

6.2.2.4 Impact on Use

Felling and Hauling Damage: Felling and hauling damage to trees left unharvested in the long term usually reduces wood quality because the injuries often lead to fungal infection with subsequent wood discoloration and decay.

In optimal cases, the wound area is discolored and the wound's fiber orientation is disrupted. This type of damage causes quality defects in veneer wood and sawn wood (Fig. 6.35). Depending on the size of the wound, tree species, season the injury occurred, and microclimatic conditions, it is possible for the decay to be so extensive that it cannot be compartmentalized. This type of damage prevents any wood of value from forming. Especially in tropical conditions, the significant amount of infection leads to the devaluation of the damaged stems.

Mechanical damage to harvested timber (stem cracks, fiber tears, rot) results in lower-quality wood or volume recovery.

The reaction of trees to injuries is already covered in Chap. 3.5 – Injury (CODIT model by Shigo 1990, (Compartmentalization of Defects in Trees), Dujesiefken and Liese 2006).

Resin Tapping: Resin tapping is not detrimental to the health and growth of pines (*Pinus sylvestris*) (Stephan 1973), but does affect the quality of wood. The stem cross section forms in the direction of the living wood.

The area on the stem around the groove is saturated with resin:

- Discoloring occurs in edge boards.
- Resin-soaked tracheids make the wood difficult to process.

Pine (*Pinus sylvestris*) tapping in late summer of the third year (single grooved) (Brandenburg, Germany, 1989)

Large callus on a pine (*Pinus sylvestris*) after 17 years, above dry crack (Lausitz, Germany)

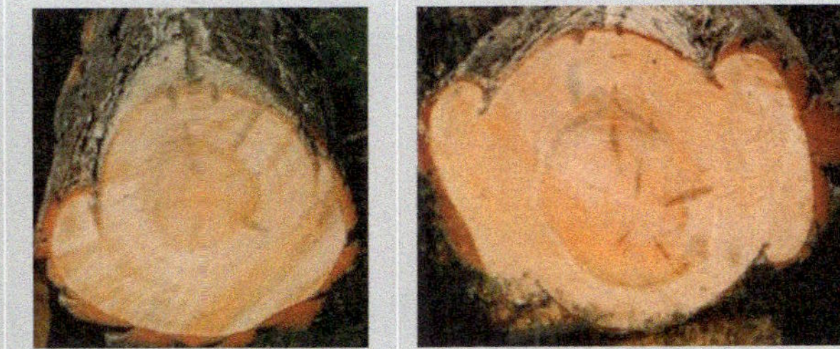

Stem deformations in two pines (*Pinus sylvestris*) with one resin groove (*left*) and two resin grooves (*right*)

Low-grade pine (*Pinus sylvestris*) board due to deformation and resin buildup

Resin tapping in an aleppo pine stand (*Pinus halepensis*) for the production of retsina wine (Peloponnese, Greece)

Barely visible diagonal notches made to extract latex on a bolletrie (*Manilkara* family) (Surinam)

Resin from a rode prokoni (*Inga alba*) is used locally for skin care (Surinam)

Resinous exudate on the base of a basralocus (*Dicorynia guianensis*) (Surinam)

Sap from hoogland mataki (*Symphonia globulifera*) is used locally to waterproof textiles and ropes (Surinam)

Plate 6.13 Use of plant exudates (forest byproduct)

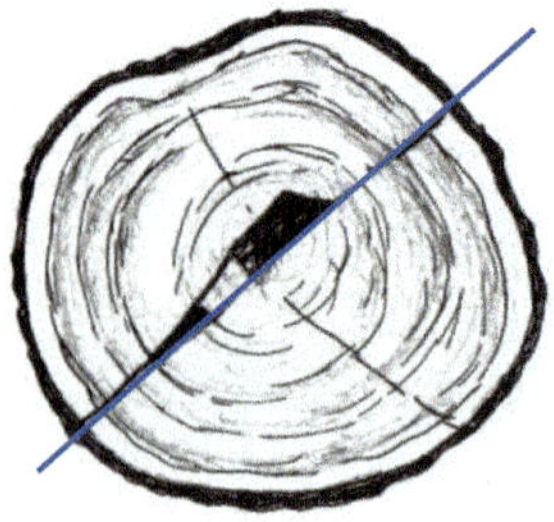

Fig. 6.36 Band saw positioning on a veneer stem with deeply split fiber

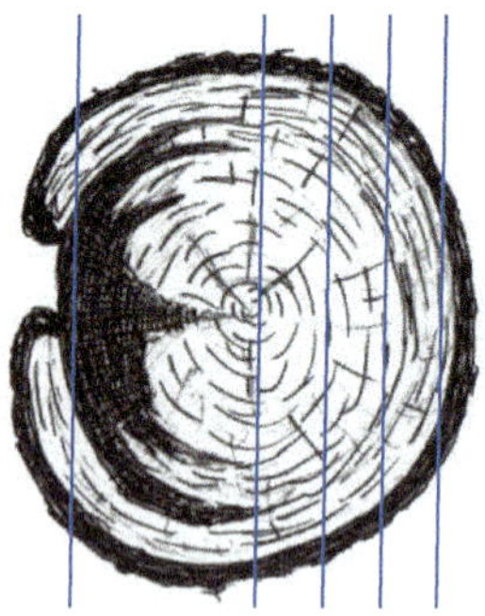

Fig. 6.37 Optimal sectioning of a spruce (*Picea abies*) stem with red rot (old hauling injury)

- Dry cracks form near the resin grooves, along the drip line.
- Wood becomes less suitable for veneer.
- Pulp and wood panels cannot have resin grooves or they will be graded as defective industrial wood.

Density, bending, and compressive strength differ after a longer compartmentalization period (approx. 20 years) significantly from untapped pine wood (Stephan 1973). A nearly 5 % loss in sawn timber volume is to be expected due to an increased amount of slab wood.

In primary and secondary forests of the tropics, trees are torn, notched, or sliced to obtain exudates (rubber, latex milk, etc.). In most cases, the trees are left standing and their injuries callused over. The hidden damage only becomes visible once the tree is logged and processed leading to a degrading of the wood.

6.2.2.5 Technological Adaptation

For wood damaged primarily down the length of the stem, the traditional methods of wood sectioning are recommended:

- Veneer logs should be cut so that the damaged area remains part of the dog board (Fig. 6.36).
- With band and gate saws, the blades should be positioned so that the damaged stem section is completely cut off (Fig. 6.37).
- With chipper canters, damaged stem sections are usually chipped away. If that is unsuccessful, then the wood should be sectioned using a circular saw log sectioning model.

Fig. 6.38 Antitank missile, grenade, and bomb fragments

6.2.3 Stem Splinters (Richter 2006b)

6.2.3.1 Description

A sawmill operator, chipboard driver, carpenter: each knows the frustration of unexpectedly finding a hidden metal object in a piece of wood. Tools break, operations are interrupted, time is wasted, and the quality of the wood plummets.

Depending on the size, shape, and type of the object, metal stem splinters can cause varying degrees of damage to a tree's bark or wood. A nail hammered into the trunk, a stray bullet from a hunter's riffle, or the many possible injuries caused by the weapons of war, all leave behind their traces. The injuries can be seen in the trees' growth deformations and uncharacteristic bark. If these are overlooked, it means trouble later on for the wood processor.

Originally, stem splinters were mainly the result of metal objects damaging trees during *warfare* (Fig. 6.38).

See Plate 6.15 for stem splinter damage inflicted during combat or hunting photos.

However, starting in the late nineteenth century, metal and wire increasingly replaced wood as the preferred material for fence building, and, as a result, more and more injuries caused by metal objects began to occur. In the mid-twentieth century, electric fences with insulators, hardened bolts, and corrosion-resistant nails expanded the arsenal of fencing material. Therefore, today, stem splinters refer to all types of metallic contamination, be it due to *carelessness or malicious behavior and vandalism* (Figs. 6.39 and 6.40), *forest activities* (Fig. 6.41), or *special interest groups* (Figs. 6.42 and 6.43).

See Plate 6.16 for forest activities including recreation, hunting, and nature conservation photos and Plate 6.17 for agricultural activities photos.

Measurement: Stem splinters are not measured. They are not even included in the relevant wood grading regulations (HKS 2002a, CEN (DIN 1997d, 1998b) and RVR 2014). However, in the official commentary on log sorting in the German states of Brandenburg and Thuringia, more attention is paid to stem splinters. In Thuringia, Germany, stem sections containing externally identifiable stem splinters "are to be extracted or sawed off." The seller typically assumes no liability for any concealed defects. "The buyer must be made

Clearing and marking the skid trails in a spruce pole wood stand (*Picea abies*) (Saxony, Germany)

The skid trails were also permanently marked for use in the succession stand. Harvest losses are accepted (Saxony, Germany)

A tropical primary forest is accessed by a logging road for a timber sale (Surinam)

Fishbone-like skid trails are laid in the jungle (Surinam)

Plate 6.14 Stand inventory

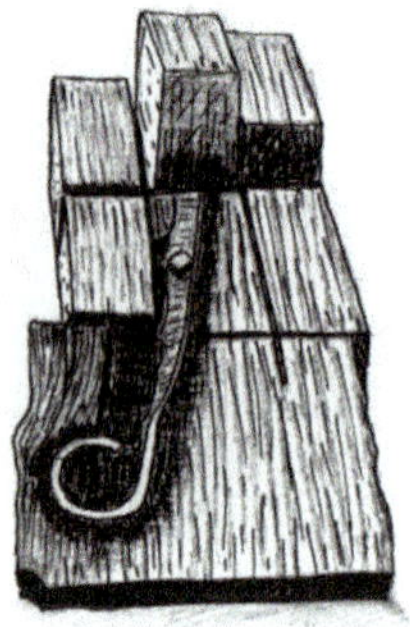

Fig. 6.39 Ingrown self-contained door hinge in oak (*Quercus* ssp.) with cutting grooves from a saw

Fig. 6.40 Ingrown tension wire and pasture fence insulators in an ash on a forest edge

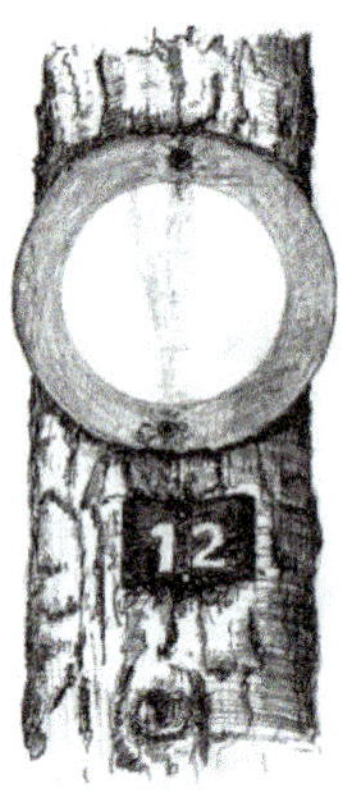

Fig. 6.41 Softwood on a forest trail with nailed in trail marker and forest service number

aware of any wood known to contain splinter wood" (Thüringer Landesforstdirektion 1997, p. 27).

This point can be countered by the argument that in general business practice, product liability applies to a concealed defect, if the defect impedes the originally intended use or if the user incurs material damage to machinery and/or tools caused by the defect. Therefore, it seems to follow that anyone selling wood should be held liable for a concealed stem splinter, particularly since in most cases, the forest owners themselves have the greatest influence on the degree of metallic contamination in their forests.

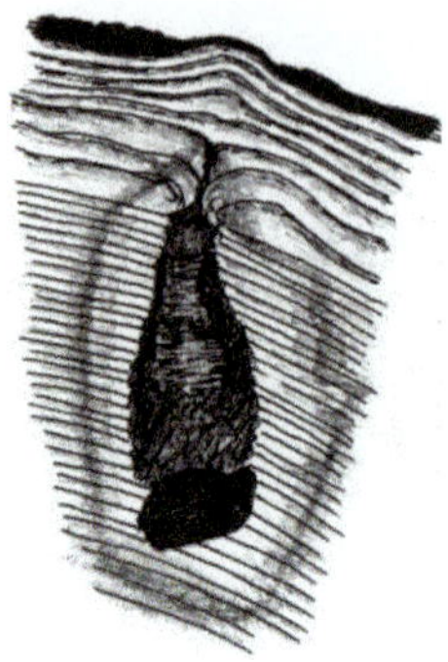

Fig. 6.42 Shot gun scar in hardwood: entry with the destroyed wood structure, oxidative discoloration, callused boarders, and bark scar

Fig. 6.43 Ingrown metal plate on an oak *(Quercus* ssp.) tree in a park

A valid addition was made to regulations in Brandenburg, Germany, stating that "wood devalued by metal objects …as with other fallen timber, can only be objectively and verifiably sorted based on established grade characteristics.…" It is prohibited to reclassify a defective piece of wood to a lower quality grade. The price must be set based on the extent of the defect (price discount) (Frommhold 2001).

Conclusion: Suspected splinter wood must be identified. Any existing metal fragments should be removed or clearly marked. When a customer agrees to purchase splinter wood, the price of the wood should reflect a discount, not a declassification to a lower quality grade.

6.2.3.2 Causes

The causes of stem splinters can be grouped into four categories:

1. Bullets and shrapnel from *warfare* and target shooting cause a large part of the deeper injuries.
2. *Carelessness* in *fence building* (pasture fences, barriers, wires, etc.), but also *malicious behavior* (using trees as shooting targets), can result in foreign objects gradually growing into the wood.
3. *Forest activities*, including *hunting* and *nature conservation*, regularly post makers or information (maps, notices, historical markers, etc.) on trees. Often the devices (nails, hooks) used to fix these signs and sometimes the signs themselves end up growing into the tree.

Oak (*Quercus* ssp.) damaged by a bomb explosion 60 years earlier (Dresden, Germany, 13 February 1945)

Oak (*Quercus* ssp.) hit by shrapnel and bullets (Bad Muskau, Germany, January 1945)

Beech (*Fagus sylvatica*) with shrapnel scars (Neisse, Germany, January 1945)

Spruce (*Picea abies*) with resin flow after being hit by a copper bullet (hunting) (*left*); wood discoloration around the wound a few years later (*right*)

Old pine (*Pinus sylvestris*) damaged by machine gunfire (Forst, Germany, January 1945) (Photo: E. Bäucker)

Alder (*Alnus glutinosa*) with splinter scar from a small-caliber projectile

Ash (*Fraxinus excelsior*) with lead splinter from a shot gun

Ash (*Fraxinus excelsior*) at a target shooting site for hunters with bullet damage

Beam with lead bullets (Liberation Wars 1812) (Rietschen, Upper Lusatia, Germany)

Plate 6.15 Stem splinter damage inflicted during combat or hunting

If the nail grows into the wood, it is destined to cause equipment damage when processed. Why not use durable tape instead (*below*)?

A gang saw broke when it hit this metal sign from the nineteenth century

Trail signs lowered the value of this maple (*Acer* ssp.)

Exemplary, individually fixed trail signs

This sign (hard metal screw and nail) was posted on this larch *(Larix decidua)* by the forest owner

This pine stem (*Pinus sylvestris*) (B quality) was used as a hunting hide. Result 7 m of splinter wood

A nesting box screwed to an oak (*Quercus robur*) trunk. Nature conservation vs.wood quality?

Spruce (*Picea abies*) embedded with the remains of a fence

The ingrown metal wires were intended to protect this maple (*Acer platanoides*) from grazing (Photo: J. Schubert)

Ingrown, heavily callused metal marker on a gronfolo (*Qualea rosea*) in a tropical rain forest test area (Surinam)

Plate 6.16 Forest activities including recreation, hunting, and nature conservation

Ingrown wire fencing in a maple (*Acer platanoides*) stimulated the formation of epicormic shoots

Ash (*Fraxinus excelsior*) on a forest edge with ingrown isolators and wire fencing

Ash (*Fraxinus excelsior*) with callused over permanent pasture boarder made from a piece of railroad track

This oak (*Quercus* ssp.) was killed by a contracted wire

Ingrown metal rope and chain used to fasten a trap on an oak (*Quercus robur*)

This hidden isolator in a linden (*Tilia* ssp.) stem destroyed a band saw blade and a chain saw

Oak (*Quercus* ssp.) log with horseshoe used as a garden fence bracket on the edge of the forest

An iron nail has greatly discolored the wood in this oak (*Quercus* ssp.) over the course of 60 years

Collection of dissected nails in a sawmill–each nail calls for a change of tools and repairs

Plate 6.17 Agricultural activities

Fig. 6.44 Properly hung trail marker with self-adhesive foil on the purified bark of an old beech (*Fagus sylvatica)*

Fig. 6.45 Hunting hide set in front, not nailed into a forked tree in a private hunting revere

4. *Special interest groups* (cultural events, tourism, recreation, sport clubs, etc.) all have information to post, events, historical markers, hiking trails, etc., and this typically involves nailing a sign on a tree.

6.2.3.3 Prevention

Conflicts must be solved through diplomacy, not war. Military force will never achieve lasting resolutions. As long as there are guns to shoot, trees will always be vulnerable to splinter injuries.

Improved and ongoing education would limit careless behavior when using trees as fixing posts (Fig. 6.44). Malicious behavior can only be curtailed through education (general acceptance of a code of conduct) or by enforcing sanctions.

Forest activities should promote better practices, for example, marking trees with paint instead of hanging up signs, using plastic nails instead of metal, or, even better, setting up separate poles for signage. Hunters and conservationists should not attach hunting and conservation objects to trees (i.e., hides, educational boards, posters, nesting boxes) (Figs. 6.45 and 6.46).

Special interest groups (cultural, tourism, sport, and recreational) should be made aware of how damaging it can be to use trees as fixtures. Where possible, the individual stakeholders should receive training through competent institutions (Anonymous 2000; DIN 2004b).

Fig. 6.46 Nature conservation keeping the tree's health in mind – the nesting box has been hung on a branch not nailed

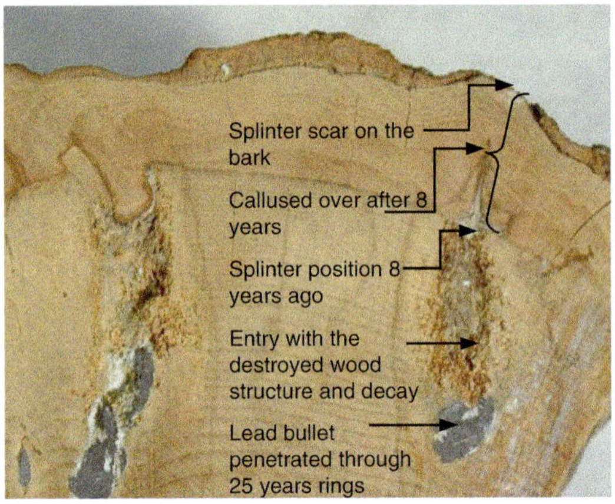

Fig. 6.47 Effects of a lead bullet injury on a hardwood

Taking care and showing respect for trees reflects a basic reference for life, "Ethics of Reverence for Life" (Schweitzer 1960).

6.2.3.4 Impact on Use

Splinter damage affects wood quality and the equipment and operational processes used in manufacturing (Vogelpohl 1987a, b).

Wood Quality: Wood substances, especially tannins, react strongly to metals. Oak (*Quercus* ssp.) and chestnut (*Castanea* ssp.) wood often turns deep blue black around the splinter. Metals are oxidized by wood moisture and atmospheric oxygen, leading to discoloration (Schute 1972a, b). The color changes in the wood alone render it almost useless for higher-value products.

Depending on the extent of damage to the wood substance, splinter wood is no longer suitable for static purposes. The wood has no potential value as pulp or chip wood because it cannot be mechanically crushed (Figs. 6.47, 6.48, and 6.49).

Fig. 6.48 Discolored wood on veneer oak (*Quercus*) caused by shrapnel, Danzig, Germany 1944 (Photo: C.T. Bues)

Fig. 6.49 Burn marks on a robinia (*Robinia pseudoacacia*) board made by a blunt saw after cutting a nail. Discolored wood in the area around the nail

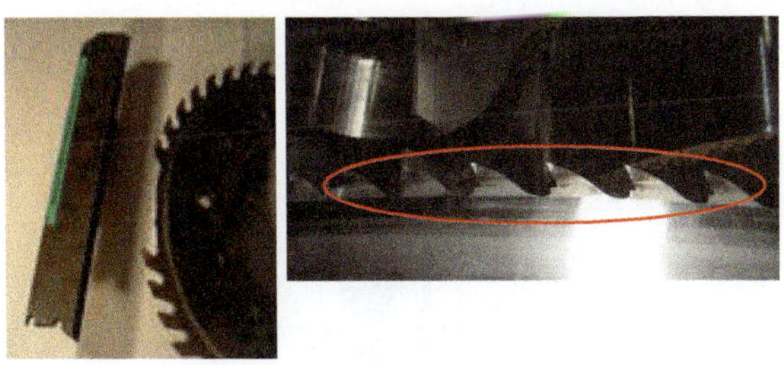

Fig. 6.50 *Left* broken blade (blade ring), broken tooth on a chipper canter. *Right*; broken teeth (*red oval line*) on a stellite band saw, caused by nails

Equipment Damage: At best, when wood processing tools hit a metallic object, it only blunts the blade. Repeated contact reduces the service life of saws, cutters, and knives. Notches mark he wood surfaces and lead to higher energy consumption.

Equipment damage leads to significant replacement costs (e.g. new band saw blades can cost € 600 [price 2012]), Fig. 6.50) and to associated machine downtime (Schute 1964).

Operational Procedures: Stem splinter damage can cause production downtime and interfere with very sensitive technological processes (Fronius 1991; Schute 1972a, b). The already low profit margins in the wood processing industry are exacerbated by incidences of stem splinter.

An extrapolation of the current cases of stem splinter (Kirbach 2008), related to 11.1 million hectares of forestland in Germany, gives the following conservative estimates (Table 6.6.).

Table 6.6 Damage by stem splinters in Germany (estimates) (Kirbach 2008)

No. of trees with stem splinter	6.4 mill.
Splinter wood volume (lower 6-m-stem sections)	4 mill. fm
No. of fixed objects	7 mill.
No. of metal fasteners	13.7 mill.
Loss in proceeds timber sales (cost-free harvest)	68.5 mill. €

6.2.3.5 Technological Adaptation

Wood workers use various diagnostic methods to locate and eliminate metallic objects before processing. The easiest method is visual inspection, especially in trees suspected of having splinters near human settlements, along roadsides and forest boundaries, as well as in forests where military battles have taken place, or on military training sites. Most metallic splinters, however, are overgrown by bark and typically do not form unusually shaped scars like branch scars. Splinter scars are characterized by a sharp edge between the old bark and the new bark that has grown over the wounded area. During callusing, a pleat forms next to the damaged wood surface. In this way, splinter scars differ significantly from branch scars found on smooth-barked species and the roses found on coarse-barked species (Fig. 6.51).

Attention! Splinter scars are not always distinguishable from scars caused by falling rocks or stem injuries caused during timber harvest.

Detecting metallic splinters with the aid of technological devices is now a common practice (Vogelpohl 1987a, b). For forest owners and small wood processing operations, *portable metal detectors* are available that can be adapted to the specific dimensions of the wood under examination (Fig. 6.52). The use of such devices requires that metallic objects be out of reach of the detection area (typically >5 m).

Companies with larger timber capacity (sawmills and chippers, particleboard and fiberboard plants, pulp and paper producers) run the debarked wood through stationary, covered metal searching and separating equipment (Fig. 6.53).

To improve their stability, clamping tools are carbide tipped (cutters, drills, saw blades) or coated in stellite (band and gang saws). However, hard metals are sensitive to impact with metal objects.

6.3 Impact of Plants: Epiphytes and Vines

6.3.1 Description (Strasburger et al. 1978; Kehl 2013; Lilienthal 2004)

Epiphytes and vines (lianas) mainly grow in tropical rain forests, but some important species are also found temperate climates. They either grow on the tree or in the ground and use the host tree for support employing various climbing techniques to spread through the crown in search of sunlight.

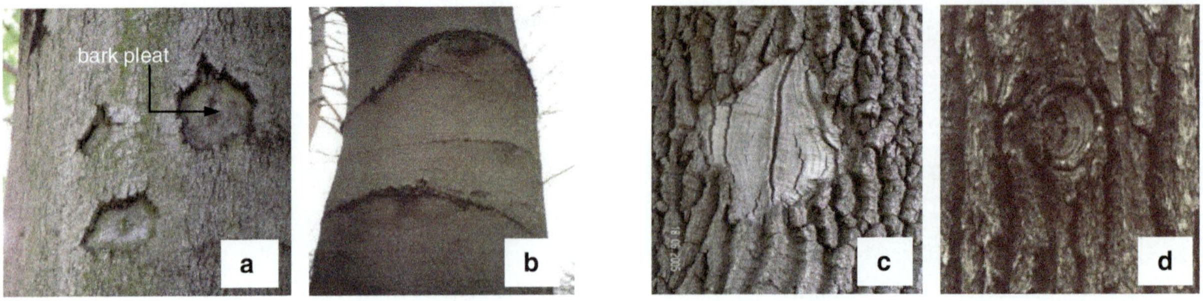

Bullet scar (a) and branch scar (Chinese´s moustache) (b) on beech Splinter scar (c) and branch scar, rose) (d) on oak (*Quercus*).

Fig. 6.51 (**a–d**) Visual difference between branch scare and 60-year-old splinter scar, Bad Muskau, Germany. Bullet scar (**a**) and branch scar (Chinese's mustache) (**b**) on beech splinter scar (**c**) and branch scar, rose) (**d**) on oak (*Quercus*)

Fig. 6.52 Using a metal detector prior to cutting a log suspected to have splinter wood

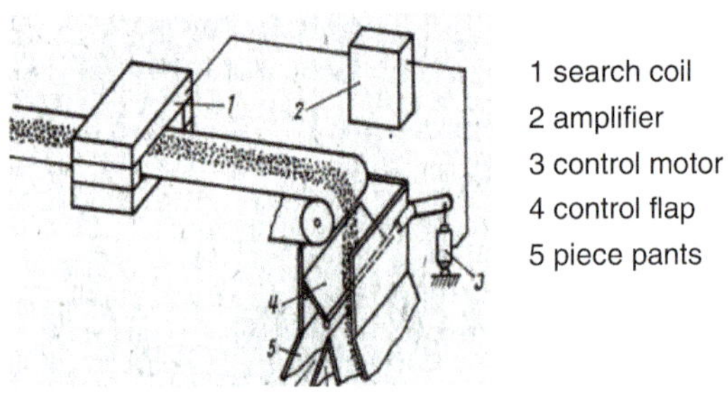

Fig. 6.53 Splinter detecting device in a belt conveyor for wood chips (Autorenkollektiv 1988). *1* search coil, *2* amplifier, *3* control motor, *4* control flap, *5* piece pants

Epiphytes grow on the host tree and rely on it for climbing support. They are widespread, from the tropics to the coastal rain forests of the temperate zone. Epiphytes germinate in tree branches, bumps, and bark in an effort to take advantage of the light penetrating the forest canopy. These include among others bromeliads (*Bromeliaceae*), orchids (*Orchidaceae*), and succulents (also cacti). Given the right growing conditions, epiphytes can become so dense they cause an entire tree to collapse under their load.

Some epiphytes live *hemiparasitically* by attaching their aerial roots (haustoria) to the branches of the host trees. In the tropics the flower plant family *Loranthaceae* is most notable. They are related to the European mistletoe (*Viscum album*), an evergreen plant commonly found growing in hardwood trees, i.e., fir (*Abies* ssp.) and pine (*Pinus* ssp.) in temperate climates. The plant's sticky white berries contain seeds which are dispersed by birds (digestive dissemination).

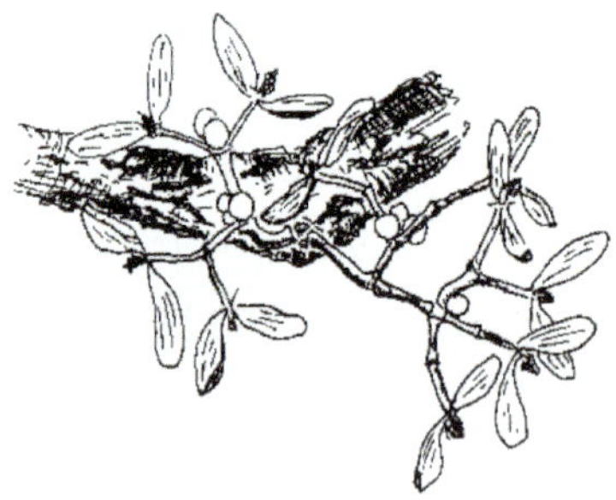

Fig. 6.54 Mistletoe (*Viscum album*) with berries

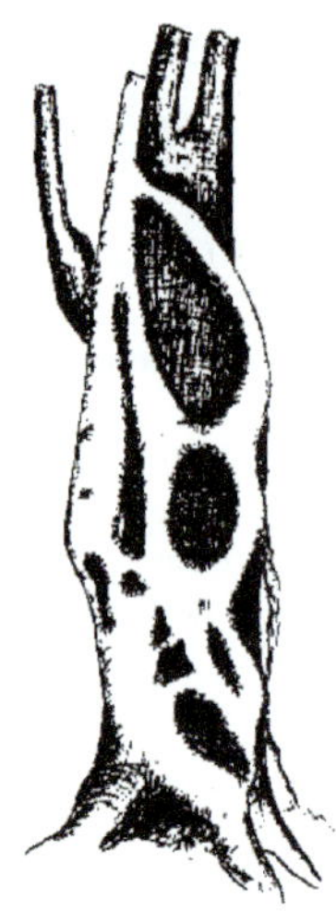

Fig. 6.55 Strangler fig (*Ficus* ssp.) encircles a host tree

They typically grow in branches. The mistletoe competes for nutrients and sunlight with the assimilation organs of the host tree (Fig. 6.54).

See Plate 6.18 for epiphytes photos.

Hemiepiphytes in the tropics are also epiphytes, whose seeds germinate from undigested bird feces on host branches. They, however, develop aerial roots which grow downwards eventually connecting to the ground. As the roots become stronger, more aerial roots develop and fuse together. The roots encircle the host tree and can eventually strangle it to death (Fig. 6.55). One notable example is the strangler fig (*Ficus* ssp.).

Left: Heavy mistletoe infestation (*Viscum album*) in a poplar plantation (*Populus tremula*) with signs of dying (Loire, France. *Upper right*: Poplar (*Populus* ssp.) with mistletoe (*Viscum album*). *Lower right*: Mistletoe haustoria in a fir branch (Tharandt, Germany)

White mistletoe (*Viscum album* ssp. *Abietis*) on a silver fir (*Abies alba*) (Tharandt, Germany)

Mediterranean cypress (*Cupressus sempervirens*) with mistletoe (*Viscum* sp.) (Evvoia, Greece)

Dwarf mistletoe (*Arceuthobium* sp.) caused witches' broom in a lodgepole pine (*Pinus contorta*) (Wyoming, USA)

Dwarf mistletoe (*Arceuthobium* sp.) caused resin flow in a Western juniper (*Juniperus occidentalis*) (Arizona, USA)

Desmaria mutabilis, a tropical flowering plant (*Loranthaceae*) (Chile) (Wikipedia, Pato Novoa, 10.1.2009)

Sitka (*Picea sitchensis*) with epiphytes in a temperate rain forest (Ucluelet, Canada)

Bromeliaceae (*Bromeliaceae*), ferns (*Polypodiopsida*), and lichen (*Lichen*) flourish during the rainy season (Surinam) (Photo: Kenter)

Epiphytes on a stem in a primary forest (Surinam)

Solitary tree enveloped by epiphytes (Surinam)

Plate 6.18 Epiphytes

Fig. 6.56 Rattan palm (*Calamus adspersus*)

Fig. 6.57 Young ivy (*Hedera helix*) vine with shade leaves

Fig. 6.58 Female wild hops (*Humulus lupulus*) with flowers

See Plate 6.19 for hemiepiphytes photos.

Vines (lianas) include over 2,500 species belonging to different plant geneses, of which approximately 90 % grow in the tropics. They germinate in the soil and use trees only for support in their rapid ascent towards the light and airy canopy. Depending on their climbing technique, they are subcategorized as scramblers, root climbers, twiners, and tendril climbers:

Tropical *scramblers* include the many species of rattan palms (*Calameae*), which use spines and prickles to spread through the host canopy (Fig. 6.56). Rich in foliage, they quickly compete for sunlight. In the temperate latitudes, scramblers include blackberry (*Rubus sectio Rubus*) and various climbing wild roses (*Rosa*).

See Plate 6.20 for vines: scramblers photos.

Root climbers develop sucker roots to help them climb up through the host tree towards the sunlight. The sucker roots do not damage the tree bark, but the plant competes for sunlight and nutrients.

In the tropics and subtropics, root climbers include among others strangler figs (*Ficus* ssp.), black pepper (*Piper nigrum*), vanilla (*Vanilla planifolia*), and climbing hydrangea (*Hydrangea petiolaris*). The main root climber in temperate forests is the evergreen ivy (*Hedera helix*). It grows with the help of sucker roots up into the tree canopy (Fig. 6.57). Ivy is shade tolerant. It can live for up to 450 years. Widespread, but not of major significance to forestry, is the wild grape (*Parthenocissus* ssp.).

See Plate 6.21 for vines: root climbers photos.

Twiners include the many species in the legume (*Fabaceae*) family. They spread their long internodes out in a circle motion (circumnutation), until they find a support to coil around and rapidly climb up to the sunlight. In temperate zones, the wild hops (*Humulus lupulus*) are particularly common. It is a left coiling annual plant typically found growing wild competing for sunlight along stand boarders (Fig. 6.58).

See Plate 6.22 for vines: twiners photos.

Tendrils are sprouts, roots, or leaves that turn into thin coiling twines. On their quest for nutrients, the tendrils seek out a climbing support. When they touch an object, they become irritated and coil themselves up (thigomonasty) then stabilize themselves again by climbing further. The wild grape (*Vitis vinifera* subsp. *sylvestris)* is one of many types of vines commonly found in the temperate climate zones.

See Plate 6.23 for vines: tendril climbers photos.

In forestry, one of the main vine species is clematis (*Clematis vitalba*). Clematis plants with their 6 cm thick stem often spread through the crowns of multiple trees (Fig. 6.59). Like other tropical lianas, clematis is able to coil so tightly around a tree that it deforms their stems and eventually strangles them to death.

6.3.2 Causes

Plants grow on trees in an effort to efficiently and effectively reach sunlight for assimilation. In the process, they cause varying degrees of damage to the host trees.

Strangler fig (*Ficus* ssp.) encircle a palm after a sinker root has reached the ground (Surinam)

A strangler fig (*Ficus* ssp.) with multiple sinker roots in the ground. The sinker roots encircle the stem. The strangler figs' foliage becomes a life-threatening competitor to the host tree (Brokopondo, Surinam)

Ficus macrophylla, planted in 1792. The sinker roots have grown into full-sized stems (Puerto de la Cruz, Tenerife, Spain)

Plate 6.19 Hemiepiphytes

Blackberry (*Rubus* sectio *Rubus*) grows stickers (Germany)

Wild rose (*Rosa fedtschenkoana*) grows thorns (Central Asia)

Bougainvillea (*Bougainvillea glabra*) grow spiky thorns (Aegean, Turkey)

Rattan palms (*Calamus reinwardtii*) grow barbed leaves and stems (Java)

Plate 6.20 Vines: scramblers

Ivy (*Hedera helix*) quickly climbs up both side of the this ash (*Fraxinus excelsior*) fork (Germany)

Self-climbing vine (*Parthenocissus quinquefolia*) (North America)

Climbing hydrangea (*Hydrangea* ssp.), Germany

Climbing hydrangea (Hydrangea ssp.) on a Douglas fir (*Pseudotsuga menziesii*) (Vancouver, Canada)

Betel (*Piper betle*) (India)

Plate 6.21 Vines: root climbers

Wild hops *(Humulus lupulus)*. The rhizomes extend annually up to 6 m in height

Wild hops *(Humulus lupulus)* with berries (Saxony-Anhalt, Germany)

Wild hops envelopes an alder (*Alnus glutinosa*). It dies after fruitification

Dutchman's pipe (*Aristolochia macrophylla*) climbs up to 20 m high

Magnolia vine *(Schisandra repanda)*, a woody climber (Japan, China)

Plate 6.22 Vines: twiners

Clematis (*Clematis vitalba*) strangles a maple (*Acer* ssp.) (Germany)

Dense network of woody clematis vines (*Clematis vitalba*) hangs from a pine (*Pinus*) (Hörselberge, Germany)

Clematis vine (*Clematis vitalba*) on an oak (*Quercus* ssp.)

Silver fir (*Abies alba*) deformed and strangled to death by a clematis vine (*Clematis vitalba*) (Niedersachsen, Germany)

Liane wind around a Douglas fir (*Pseudotsuga menziesii*) (Vancouver Canada)

Wild grape (*Vitis vinifera* subsp. *sylvestris*) envelopes an ash (*Fraxinus velutina*) (Zion National Park, Utah, USA)

Chinese wisteria (*Wisteria sinensis*) winds around a support post (East Asia)

Network of jungle vines (Surinam) (Photo: Kenter)

Liane tightens its trunk between host tree and the ground (Surinam)

Plate 6.23 Vines: tendril climbers

Liane *(Bauhinia guianensis)* consisting of partly wavy, tangled stem parts

Major tree species such as gronfolo (*Qualea rosea*) (*left*), the valuable bolletrie stem (*Manilkara bidentata*) (*center*), or poteria (*Pouteria sagotiana*), a tree with pharmaceutical properties, are all equally entwined by various liane plants (Surinam)

After logging a green carpet of lianas covers the forest floor (Surinam)

A vine strangling parts of tree (Surinam)

This liane interfered with the host's assimilation causing the stem to thicken (Surinam)

The vine deformed the stem then died. The stem cannot grow out of the deformation (Surinam) (Photo: Kenter)

Marks left by a vine suggesting drape textured figure (Surinam) (Photo: Kenter)

Sapeli drape veneer (*Entandro phragma cylindricum*), caused by a vine

Plate 6.23 (continued)

Fig. 6.59 Clematis (*Clematis vitalba*) with multiple fruit

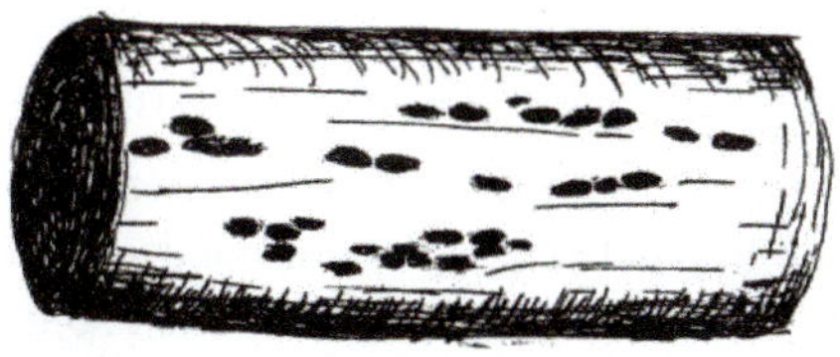

Fig. 6.60 Traces of mistletoe (*Viscum album* ssp. *Abietis*) roots

Epiphytes affect their host as they climb by creating their own growing environments of humus and water retention which can increase the host tree's risk of infection. They also block the tree's leaves from sunlight. Hemiparasitic climbers are even more invasive. They use their haustorial roots to draw nourishment and water from the host tree (Fig. 6.60) and compete for sunlight in the crown area. Heavy mistletoe infestation can kill a host tree. Some (dwarf) mistletoe species (*Arceuthobium* sp.) target specific host trees causing a proliferation of witches' broom in, for example, North American pine species (*Pinus* ssp.), junipers (*Juniperus* ssp.), Douglas fir (*Pseudotsuga menziesii*), and cypress (*Cupressus* ssp.).

The growth "strategy" of *hemiepiphytes* is to use the host tree for support until its own network of aerial roots becomes self-sustaining. If the host tree dies in the process, it becomes the nutrient supplier for the hemiepiphyte.

Vines (lianes) which germinate in the soil use thorns, spines, sucker roots, tendrils, and coiling motions to climb up the host tree towards the sun.

While the scramblers, root climbers, and twiners do not usually cause damage to the host tree, other than competing for light and nourishment, tendril climbers impact the tree's cambial growth. The suffocating tendrils do not grow along with the tree but rather constrict growth by coiling around the stem.

Lianes such as rattan (*Calameae*) can significantly impact a host tree's growth. They can become so dense, enveloping the host canopy with up to 300 m long vines that they quickly suffocate the tree from sunlight. In warmer regions of the temperate zones, wild grape and to a lesser extent the wild hops are particularly influential (Fig. 6.61).

Fig. 6.61 Maple (*Acer* ssp.) enveloped by clematis vines

6.3.3 Prevention

Epiphytes and *hemiepiphytes* are nearly impossible to prevent; doing so would require either eliminating all the birds that act as seed carriers or mechanically removing all the air plants.

Vines (lianes) can theoretically be cut off at the stem base. In principle, the question should always be asked whether the presence of this species in an intact biogeographical region is not of higher value than the possible damage to the affected trees.

6.3.4 Impact on Use

The hemiparasites among the epiphytes (mistletoe (*Viscaceae*)) typically only impact the wood structure of the host tree's branches and thus do not threaten the stem. Heavy mistletoe infestations, however, can be extremely damaging if while competing for nutrients, they end up killing the host tree (poplar plantations (*Populus* ssp.)), fruit and park trees, and entire softwood stands (USA, CAN)). Mistletoes are highly sought after as holiday decorations.

Studies show mistletoe can be used as an indicator for heavy metal pollution in soils. Heavy metals reduce the resistance of mistletoe infestation among poplars (N N 2009a).

Vines can wrap so tightly around a host tree that they deform the stem and ultimately kill it. For this reason, vines are usually removed from commercial forests.

With wild hops (*Humulus lupulus*), the question is whether its negative impact, particularly along forest boarders, outweighs its attractive appearance. It also contains psychoactive substances that are pharmaceutically interesting as a hemp crop.

Ivy (*Hedera helix*) in trees is a welcome "decoration." It blooms late and is an important honey plant; the fruits are consumed by many different bird species. Ivy leaves contain saponin, a drug used to sooth bronchitis.

Tightly coiling tropical vines can also change the shape of a stem so dramatically that it creates beautiful textures that appear when the wood is processed into veneer. This is referred to as drape veneer (Fig. 6.62).

Fig. 6.62 Sapeli drape veneer (*Entandrophragma cylindricum*)

6.3.5 Technological Adaptation

In commercial forests, trees with heavy mistletoe infestation should be harvested before they die. During harvest operations in tropical primary and secondary forests, vines should be cut off before the trees are felled so that they do not damage or even pull out other trees when they fall.

7 Abiotically Induced Wood Characteristics

7.1 Temperature-Humidity Effect

7.1.1 Bark Scorch/Sunburn (Richter 2006c)

7.1.1.1 Description

If you look carefully at young city trees planted without sun protection, you will often find cracks in their stems on the side most exposed to the sun after the first growth season. This is an example of cost-saving measures with disastrous consequence.

The bark of thin-barked trees such as beech, hornbeam, sycamore, maple, lime, spruce, and the less common fir and white pine (*Fagus sylvatica*, *Carpinus betulus*, *Acer platanoides*, *Acer pseudoplatanus*, *Tilia cordata*, *Picea abies*, *Abies alba*, *Pinus strobus*) is especially prone to crack under strong solar radiation, in either long strips or patches (Fig. 7.1). The dying bark changes color, peels away from the trunk, and subsequently exposes the wood tissue. Wood discoloration and fungal and insect infestations reduce the wood's stability and value.

Attention! This damaging abiotic factor should not be confused with the symptoms of bark disease caused by fungus infection, such as the poplar bark disease caused by *Cryptodiaporthe populea*. See Plate 7.1 for Bark scorch/sunburn photos.

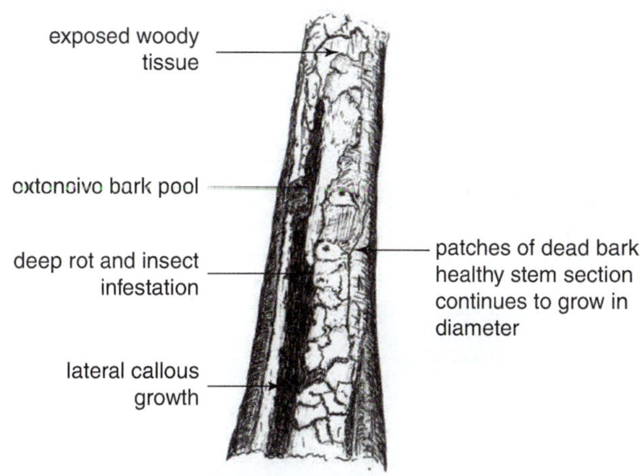

Fig. 7.1 Scorch damage on a mature beech (*Fagus sylvatica*)

C. Richter, *Wood Characteristics: Description, Causes, Prevention, Impact on Use and Technological Adaptation*,
DOI 10.1007/978-3-319-07422-1_7, © Springer International Publishing Switzerland 2015

20 years after sudden SE exposure, the trees in this beech (*Fagus sylvatica*) stand are either dead or heavily scorched

This beech (*Fagus sylvatica*) reacted to sudden exposure by forming an extra layer of bark

Dry patches of bark have callused over in this beech (*Fagus sylvatica*)

Beech (*Fagus sylvatica*) with extensive rot caused by bark scorch.

Old beech (*Fagus sylvatica*) trees which died after sudden exposure

Strip logging with SO-exposed spruce (*Picea abies*) quickly shows signs of bark scorch (*left*). 30 years of bark scorch has devalued all the boarder trees in the spruce (*Picea abies*) stand (Saxony, Germany)

Bark-scorched spruce (*Picea abies*) 20 years after sudden exposure

This spruce (*Picea abies*) protected itself against scorching by forming scaly bark

Linden (*Tilia cordata*) without sun protection damaged by bark scorch (sunburned bark and callus formation)

Plate 7.1 Bark scorch/sunburn

A roadside planting of hornbeam (*Carpinus betulus*) (*left*) and lime (*Tlia cordata*) (*center*) without sun protections suffered serious radiation damage. The bark scorch is not yet healed after 12 years. After 7 years a replacement planting is necessary–this time using sun protection tube mats (*right*)

The trunk of a suddenly exposed park linden is protected from bark scorch with reed matting (Sanssouci, Deutschland)

A giant tree falls in the jungle leaving a large hole in the canopy which leads to sunburns on the exposed trees (Surinam)

This suddenly exposed zwarte pintolokus (*Talisia pedicellaris*) in the tropical rain forest is a candidate for sun burn (Surinam)

Plate 7.1 (continued)

Figs. 7.2 and 7.3 Partially callused bark on a mature spruce (*Picea abies*) on a felling face (Fig. 7.2, *left*) and winter scorch on a young linden (*Tilia* ssp.) (roadside planting) (Fig. 7.3, *right*)

7.1.1.2 Causes

When trees with thin bark are suddenly exposed to the sun on the south or southwest side, the intense summer sun can scorch the bark and dry out the cambium until it ultimately dies. Typically, such sudden exposure is the result of an opening caused by the loss of neighboring tree branches due to wind or snow or openings cut for road and power lines. But bark scorch can also be the result of poor forest management practices, such as logging forests from the south to southwest, removing boarder trees, suddenly exposing trees to sunlight which for decades had been protected from sunlight, as well as neglecting proper spacing requirements at planting or during stand maintenance (Fig. 7.2).

On winter days with high sun radiation, bark and cambium can dry out due to insufficient water supply. While evaporation takes place in the crown area, ground frost prevents the tree from absorbing any water. Shallow rooted trees (spruce) and young (urban) trees (Fig. 7.3) are particularly vulnerable.

Many tropical tree species typically grow in dense primary forests. There bark does not need protection from direct sunlight. The trees use their limited recourses primarily to expand their crowns not to grow thicker bark. When these thin-barked trees are suddenly exposed, it can lead to sunburns.

7.1.1.3 Prevention

Openings in sunlight-exposed directions (felling face, gap cuts, road lines) are to be avoided. A protective boarder of trees must be maintained. If an abrupt boarder occurs due to natural events (wind throw, snow pressure), then any trees damaged by bark scorch should be left to shield the remaining trees from further damage.

Stand boarders should consist primarily of coarse-barked trees and shrubs. Openings should be immediately replanted so that exposed trees are shaded again.

Fig. 7.4 Correct selection of tree species for a stand boarder providing stem protection and shade to protect against scorching

Free standing trees (urban trees) require some form of sun protection when they are first planted to prevent burns and evaporation (Fig. 7.4).

7.1.1.4 Impact on Use

As the amount of damage to the bark increases, the value of the wood decreases. Stem dryness and fungal and insect infestation prevent any higher use and can lead to degrading of the wood to D quality.

Bark burn on young trees is usually callused over, but later often leads to heart rot in the stem. In urban areas such trees are often cut down for safety reasons.

7.1.1.5 Technological Adaptation

Undamaged stem parts can be strategically cut for further processing (Fig. 7.5). This, however, is seldom the case because it either requires too much effort or due to a lack of skills.

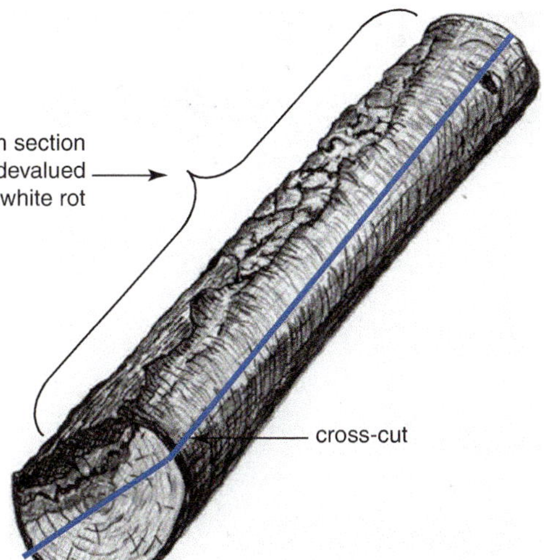

Fig. 7.5 Suggested processing for a beech (*Fagus sylvatica*) log damaged by bark scorch

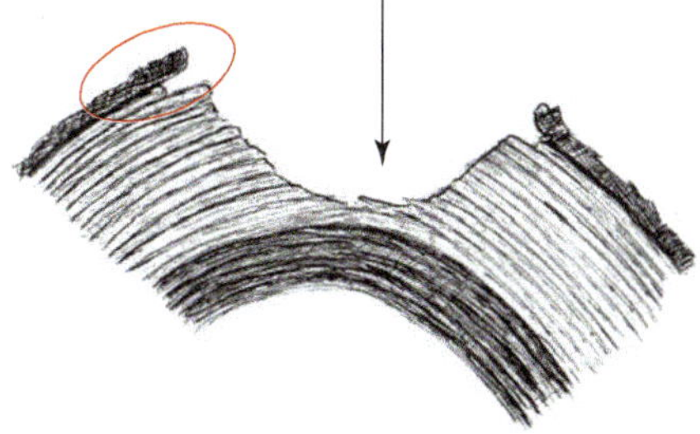

Fig. 7.7 Cypress (*Chamaecyparis lawsoniana*) stem cross section with lightning groove (average strength strike)

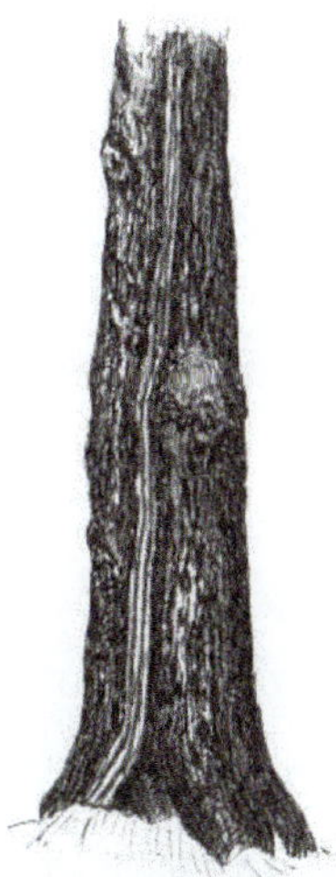

Fig. 7.6 Ash (*Fraxinus excelsior*) with callused lightning shake (weak lightning strike)

Fig. 7.8 Burst open spruce (*Picea abies*) stem (strong strike)

Fig. 7.9 Lightning hole with dying trees in a spruce forest caused by diffused lightning

7.1.2 Lightning: Lightning Groove, Lightning Hole (Richter and Bues 2005)

7.1.2.1 Description

In temperate climates, trees that have been struck by lightning are a common sight. The consequences of the frequent lightning strikes depend on the type of bark (smooth or coarse), wetness of the bark surface, and intensity of the lighting.

Lightning can strike a rain-soaked, smooth-barked tree and flow out into the ground without causing any visible damage. When it strikes a rough-barked tree, it often leaves a narrow channel down the length of the stem called a *lightning groove*. The cambium immediately surrounding the lightning shake eventually dies and the bark slowly peels away. If the tree is still vital, it will try to callus over the injury with wound tissue (Fig. 7.6).

A powerful lightning strike can blast off strips of bark or completely shatter the tree to pieces (Figs. 7.7 and 7.8). The destruction caused by lightning is impressive and a bold demonstration of the awesome power of nature.

A *lightning hole* refers to a group of trees in a softwood stand that were damaged by lightning and slowly died. Their needles changed color and any branches protruding into the center of the lightning hole turn brown and fall off. Mistakenly, this occurrence is often attributed to bark beetle disease, although the bark beetle infestation actually follows as a result of the lightning strike and the resulting loss in

Cloud to cloud lightning (schematic diagram)

Lightning striking a tree (schematic diagram)

Powerful lightning strike destroyed a beech (*Fagus sylvatica*) and splits a second in half (photo: Archive Chair of Forest Utilization Tharandt)

Very heavy lightning strike damaged this spruce (*Picea abies*) (Tharandt, Germany)

Large lightning groove in a maple (*Acer pseudoplatanus*). Typically most damage is the lower part of the stem (Germany)

The lightning channel always follows the fiber course; in this oak (*Quercus* ssp.) it is right spiraling (Wörlitz, Germany)

Two recent lightning grooves in an oak stem (*Quercus* ssp.) (Wörlitz, Germany)

Lightning groove with lateral callusing in an oak (*Quercus* ssp.) drying for three years (Photo: J. Schubert)

Plate 7.2 Lightning shakes and holes

Callused lightning groove in a larch (*Larix*), after a low-intensity lightning strike (Tharandt, Germany)

Thuja (*Thuja occidentalis*) with lighting groove. St-ong lightning strike (li.). The weaker lightning groove on the other opposite stem side was almost completely healed in 10 years (Tharandt, Germany)

Spruce (*Picea abies*) hit by strong lateral lightning strike. Greatest damage to lower part of the stem (Oberhof, Germany)

Diffused lightning discharge, also cause of lightning holes in forests (schematic diagram)

Large-scale diffused discharge led to lightning hole (stem and crown damage) (Bavaria, Germany) (Photo: C.T. Bues)

Spruce (*Picea abies*) crowns damaged by diffused lightning discharge, (Bavaria, Germany) (Photo: C.T. Bues)

Plate 7.2 (continued)

vitality of the affected trees. The largest recorded lightning hole to date is 3,500 m², the smallest is 200 m² (Fig. 7.9) (Heß-Beck 1927/1930).

See Plate 7.2 for lightning shakes and holes photos.

7.1.2.2 Causes

In temperate climates lightning is a quick, very bright flash of light caused by lightning either within the storm clouds piled up between 800 m and even 12,000 m altitude or between the clouds and the ground. Depending on sign of the voltage difference were differentiated between upwards or downwards directed lightning (Anonymous 2005). The voltage differences can amount to some 100 million Volts, with currents of several 100,000 Amps. For a millionth of a second after the lighting flash, the air immediately heats up to above 30,000 °C. The diameter of a flash can be a few centimeters to several meters (Anonymous 2005). The sudden expansion of air in the lightning channel subsequently causes the thunderclap. Despite the brilliant flash of light and booming thunder, the amount of energy released is about the same amount of energy contained in 10 l of fuel oil (Aaftrink et al. 2005).

There are three types of lightning: The first is created by a discharge within a cloud, appearing as a bluish glow around cloud's perimeter.

The second type occurs when there is a discharge between a cloud and a pocket of air in the surrounding sky (cloud-cloud lightning). The result can be a lightning bolt up to 60 km long. The direction of the flash is determined almost exclusively by the distribution of electrical charges in the atmosphere. Exposed points at ground level, such as treetops, can also be strike points.

The third type is created by a discharge between a cloud and the ground (cloud-ground lightning). Only about 10 % of all lightning discharges take place between a thunderstorm cell (lower part of the cell is negatively charged) and the ground (positive charge). Each discharge is preceded by a leader stroke which travels from the cloud to the ground. The leader stroke is very weak (1–10 A) with negative polarity and progresses along a brightly lit zigzag channel with frequent side branches. As soon as it reaches the ground, there is a stronger return stroke back to the cloud (possibly several in succession) which has a positive charge and a voltage of 10^4–10^5 A.

When a leader stroke hits a tree, the main discharge follows. If the tree has a resistance of $10^4\,\Omega$/m, an electrical current with voltages between 10^5 and 10^6 V/m will travel down the length of the stem resulting in a flashover. The tree's rough bark presents resistance to the flashover, so the current travels down the stem, through the wood (Baatz 1985), following the course of the fiber. In its path, the lightning channel boils the water contained just beneath the bark such that it explodes in a kind of steam shock effect. The extent of injury thus depends on the texture of the bark and the strength of the lightning. Trees with smooth bark can escape without significant damage. Typically, the sapwood is damaged as the lightning channel carves out a scar along the length of the stem.

Attention! A so-called lightning "shake" only develops when the exposed stem wood dries out below the fiber saturation point.

In forestry practice, lightning holes are attributed to the effects of diffused lightning or sheet lightning. Sheet lightning occurs when the individual discharges of a lightning strike are broken up by the spatial distribution and rapid movement of air masses and then discharged over a wide area of forestland. In such cases, not only does the diffused lightning kill any trees directly hit, but the surge of voltage spreads through the surrounding ground damaging and eventually killing neighboring trees as well (Gäbler 1962).

The radial voltage gradient can be so extreme that it injures any humans and animals caught near the strike point. The size of the voltage gradient depends on the intensity of the electricity discharged into the ground as well as the specific structure and conductivity of the soil substrate. The size of the voltage gradient largely determines the area of the lightning hole (Fig. 7.10).

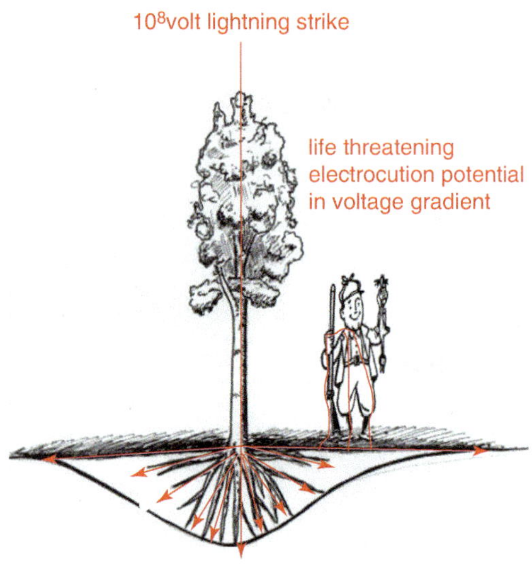

Fig. 7.10 Voltage gradient and the potential area of electrocution danger after a lightning strike

In meteorology, sheet lightning refers to remote intracloud lightning that appears as a diffused glow on the surface of the clouds very far from the spectator (lightning) (Berger 1966). The actual lightning channel is hidden by particles of precipitation and dense clouds (Anonymous 2003). Such sheet lightning is therefore unlikely to create lightning holes.

In the tropics, cloud to ground lightning is rare, during thunderstorms, because the lowest boundary of the storm clouds is usually at an altitude of about 3,000 m. Lightning strikes from 3,000 m to the ground are almost impossible (DWD 2007).

7.1.2.3 Prevention

Lightning cannot be prevented. Thunderstorm activity and lightning frequency varies regionally. For Bavaria (Germany), 5,010 flashes (3,500–17,990) per square kilometer per year are registered on average (Bosshard 1984a, p. 234). However, it is only 0.5–10 impacts per km^2. In Germany, around two million lightning strikes were registered in 2003.

Worldwide as many as 2,000 people are killed each year by lightning; many more are injured (Anonymous 2003). As lightning is unpredictable and practically every tree with almost equal probability can be struck, seeking shelter during storms under trees can be fatal. It is best to stay at least several meters away from any tall object and to crouch down with as little contact to the ground as possible so that if lightning does strike, the current can pass through with as little resistance as possible. The old saying "beware of an oak, it draws the stroke; avoid an ash, it counts the flash; creep under the thorn, it can save you from harm" can be fatally incorrect. Although some tree species are better at conducting lightning because of their smooth bark, the amount of electricity flowing into the ground around a tree hit by lightning remains the same regardless of the species.

Attention! It is also extremely dangerous to follow the alleged "moon rules," which derive a nonsensical link between trees with left spiral and lightning strikes: "in addition, lightning only strikes trees with left spiraling grain – useful information if caught in a forest during a thunderstorm. Only stand under straight grained or right spiral trees" (Paungger and Poppe 2000, p. 175). This is completely false advice and can prove deadly (see photo of a right-hand spiral oak struck by lightning in the photo section!). Buildings or cars are good places to seek protection because the metal exterior functions as a Faraday cage.

7.1.2.4 Impact on Use

Damage to the wood usually occurs only directly along the lightning groove in the water-conducting sapwood. The wood fibers a few millimeters beneath the groove typically remain undamaged. But because the parts of the stem without bark quickly dry below the fiber saturation point (approximately 30 % moisture content), dry cracks form creating entryways for insects and fungi. Woodpecker holes indicate further stem damage (Fig. 7.11).

Research done after a lightning event that resulted in the slow death of 153 ninety-year-old spruce trees on a 0.23 ha plot showed no significant difference between affected and unaffected trees in their wood moisture content, neither in the gross density-dependent percentage of water content, the gross density, or the compressive strength (Bues et al. 1989).

There is no influence on the static resistance of the stem wood sections not directly damaged by the lightning. Recent lightning damage should therefore only lead to reduction in price or a diameter compensation if the stem surface is damaged beyond the portion of slab wood. This could be the case, for example, if the lightning groove follows along the spiral grain of a stem's valuable butt end.

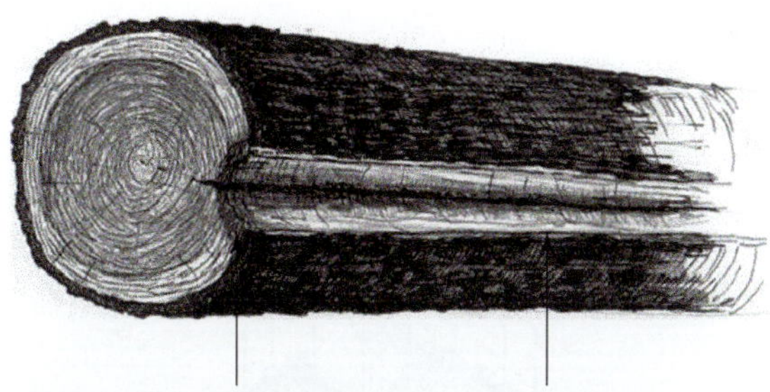

Fig. 7.11 Devalued wood from an oak (*Quercus* ssp.) stem with an old lightning groove

7.1.2.5 Technological Adaptation

At the sawmill lightning damaged logs can be processed using a rift saw, whereby the damaged part of the stem is cut off as slab wood with the first cut (Fig. 7.12). In plain sawn timber, the lightning groove can be located in the narrow edge of the board, so that the injured area can be cut off later as slab wood (Fig. 7.13).

7.1.3 Frost Cracks, Frost Scars (Richter 2002d)

7.1.3.1 Description

It sounds like a gunshot when on a frosty winter day, an ash (*Fraxinus excelsior*) trunk suddenly ruptures sending a deep, long crack up the length of its trunk. Sometimes these gaps are so wide you can fit your finger in them. Such *frost cracks*, also called freeze cracks, are fissures on the outer bark of a tree that typically run with the course of the fiber and often penetrate deep into the tree's core. They usually start at the stem base and can run up into the higher sections of the trunk (Fig. 7.14).

Frost scars, also called duck bills, refer to layers of callus tissue that form over a repeatedly cracked frost injury (Fig. 7.15). Frost cracks occur most often in solid hardwoods such as oak (*Quercus*), ash (*Fraxinus excelsior*), elm (*Ulmus* ssp.), maple (*Acer* ssp.), and chestnut (*Castanea sativa*), less frequently in soft hardwoods, with the exception of poplar (*Populus* ssp.), and more infrequently in softwoods (*Coniferales*).

Measurement: Frost cracks and frost scars are usually not measured, but are always taken into consideration when determining wood quality. The crack's age, depth of penetration, and length are recorded as well as the width and height of the protruding frost scar.

See Plate 7.3 for frost cracks, frost scars photos.

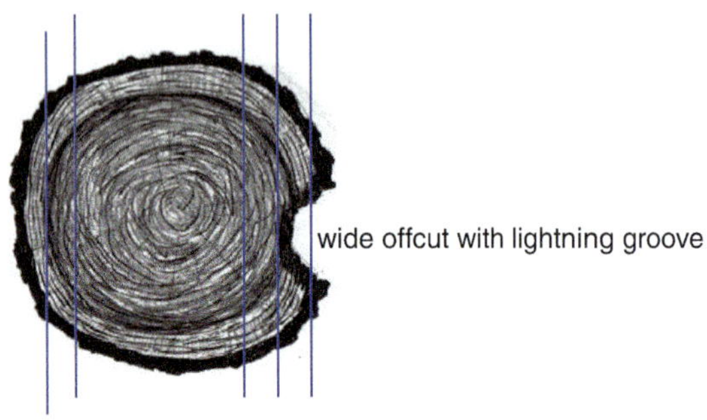

Fig. 7.12 Rift sawn log with lightning groove

Fig. 7.14 Ash (*Fraxinus excelsior*) with repeatedly split frost scar

Fig. 7.13 Plain sawn log with lightning groove

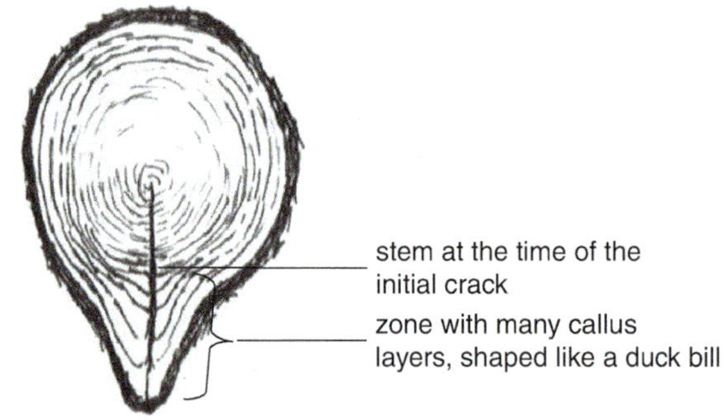

Fig. 7.15 Stem cross section with frost crack and frost scar

Blue skies following a low-pressure area, abrupt drop in temperature, radiation frost– conditions favorable to frost cracks (Oberpfalz, Germany)

Fresh frost crack on an oak (*Quercus* ssp.)

Ash (*Fraxinus excelsior*) with a re-split frost crack at –15°C

Severe frost scar following the fiber course of an oak stem (*Quercus* ssp.)

Extreme "duck bill" on an oak (*Quercus* ssp.)

Frost scar on the initial crack in an oak (*Quercus* ssp.), braced (unfortunately with iron clips – leading to tannin stains)

Frost cracks on the base of oak (*Quercus* ssp.) stems likely started by bacteria in the heartwood then supported by frost events to extend out to the stem surface as frost scars (Schmidt et al. 2001). *Far left*: fresh crack appearing on the stem surface; *left*: several older frost cracks have reached the stem surface; *right*: fresh crack which will close up again during the vegetation period; *far right*: oak cross section with frost crack

Plate 7.3 Frost cracks, frost scars

Beech (*Fagus sylvatica*) with faint, high reaching, young frost crack	Old beech (*Fagus sylvatica*) with long, prominent, frost scar	Birch (*Betula pendula*) with prominent frost scar starting 6 m up the stem	Norway maple and sycamore maple (*Acer platanoides, Acer pseudoplatanus*) with frost scars. The crack starts in the deformed heartwood and extends to the outer bark

Ash (*Fraxinus excelsior*) with gaping frost crack running all the way to the pith	Umbrella tree with prominent frost scars (*Magnolia tripetala*) (Tharandt, Germany)	Coastal redwood (*Sequoia sempervirens*) with frost scars at 4 m height (Sopron, Hungry)	Oak log (*Quercus* ssp.) for sale with a large frost scar (Saxony, Germany)

Plate 7.3 (continued)

7.1.3.2 Causes

Frost cracks occur when temperatures suddenly drop below the freezing point. When this occurs, the stem experiences a type of cold shock. The temperature gradient creates an unequal thermal contraction of the cells between the cold stem surface and warmer core resulting in the crack (Kollmann 1942). The thermal contraction alone should not be overvalued because the thermal expansion of the fiber in the longitudinal direction only amounts to 0.00 00 03 16 mm/°C, in radial direction the 9 times (Bues et al. 2005). So, for example, a temperature drop of 15 °C would only cause the stem surface of a 2 m wide tree to shrink by approx. 0.84 mm.

To a lesser extent, the radial-tangential anisotropy of the wood reinforces the tension in the wood as the stem's exterior cools. The wood contracts tangential to radial in a ratio of nearly 1.5 (Niemz 1993).

A main cause of cracking is the dehydrating effect of the freeze. Moisture is withdrawn from the cell walls of the cell lumen in the form of ice causing the cell walls to shrink. This also takes place tangentially more than radial (Kübler 1962; Schirp 1967) (Fig. 7.16).

The cracks typically initiate on vulnerable spots in the xylem (Dujesiefken 2001). These can be old bark injuries and abnormalities in the stem surface and then spread out from there. In other words, the underlying cause of the cracking is some sort of weak area on the bark, and the trigger is the sudden cooling of the stem surface below the freezing point. The susceptibility of cracking is increased by high moisture content, straight grain, large vessels, large wood rays, and a large proportion of parenchyma (Wagenführ and Scheiber 1989).

Frost scars are the result of an unsuccessful attempt by the tree to callus over a frost crack. Frost scars protrude significantly from the stem surface.

Attention! Research on oaks with externally visible cracks at their stem base has shown that *bacteria* can also cause radial shakes in living trees. These cracks often run from the heartwood to the sapwood and eventually end up as so-called frost scars (Schmidt et al. 2001). Presumably the breakup of the wood structure (breakdown of parenchyma, pectin and cellulose derivatives) causes the cracks to widen further.

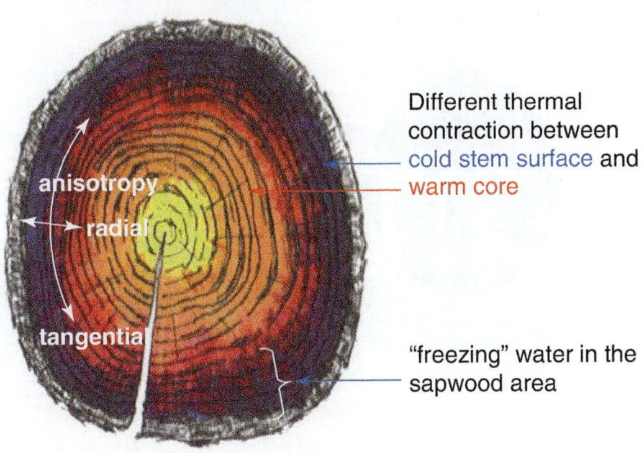

Fig. 7.16 Tension distribution on a freshly cracked stem

7.1.3.3 Prevention

Factors inducing cracks should be mitigated, as far as possible. The most vulnerable trees grow on stand edges, on valley slopes, and in ravines susceptible to cold air flows and sites subject to cold air pockets. Oak (*Quercus* ssp.) and maple (*Acer* ssp.) are particularly prone to cracking with wide wood rays serving as a potential fissure channels. In addition to choosing appropriate tree species, frost cracks can also be avoided by improving site conditions through sound forest management practices which protect against frost pockets and cold air streams (Fig. 7.17).

7.1.3.4 Impact on Use

Because frost cracks can extend along the full length of the stem and are often fairly deep, they are typically considered major defects that exclude the wood from being graded as high quality (DIN 1999b; HKS 2002a). Cracks in spiral grain trees which follow the course of the fiber and emerge on the working surface of a log significantly impact the usefulness of the wood (Fig. 7.18).

Crucial elements of a wood quality classification, apart from quality of the rest of the stem, are the age of the frost

Fig. 7.17 Trees with frost cracks in a creek bed. Cold air surges caused repeated frost cracks on the exposed side of the stem

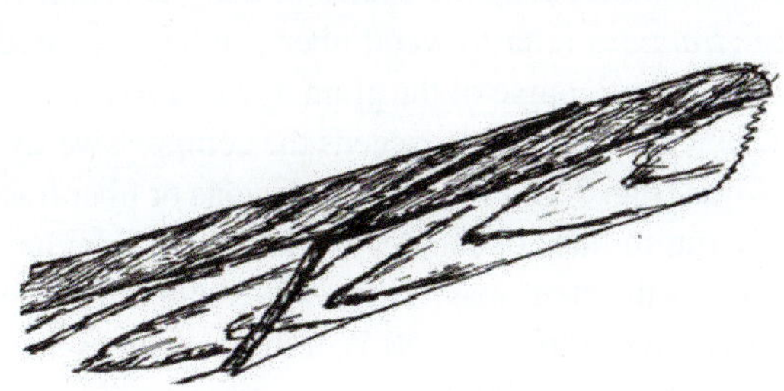

Fig. 7.18 Frost crack markings on a board from a spiral grain stem

Table 7.1 Impact of frost cracks/frost scars on quality grade set forth in RVR 2014

Species		RVR quality grade			
		A	B	C	D
All tree species except oak (*Quercus* ssp.)		No quantitative information			
Oak	Frost crack	Inadmissible	Inadmissible	Admissible if frost crack is straight	Admissible

crack and the associated risk of oxidative discoloration around the cracked area and fungus attack. The age of a frost crack in a log can be determined by counting the number of tree rings formed by callus layers of the frost scar.

The special features of frost cracks are interpreted differently in various grading standards. An obvious "gray area" exists in the different standards applied by the individual states of Germany. The CEN standards only specify grading rules for frost cracks in oak (*Quercus* ssp.) and poplar (*Populus* ssp.) (DIN 1997d; DIN 1998a; DIN 1998b; DIN 1998c); among the other species there is room left for interpretation. In general, the European standards for grading lumber with frost cracks are stricter than the German regulations (HKS 2002a). In the Framework Agreement for Timber Trade in Germany (RVR), only frost cracks in oak (*Quercus* ssp.) are relevant (RVR 2014) (Table 7.1).

7.1.3.5 Technological Adaptation

Frost cracks are not suitable for rotary cut veneer because the veneer ribbon would split after each rotation of the stem. Valuable straight stem sections with an axis-parallel frost crack are suitable as sliced veneer, if the stem is cut before in two pieces with a band saw at the right angle to the crack (Kollmann 1942) (Fig. 7.19). If the crack is located in the direction of the cut in a saw log, then it is possible to achieve a good yield (Fig. 7.20).

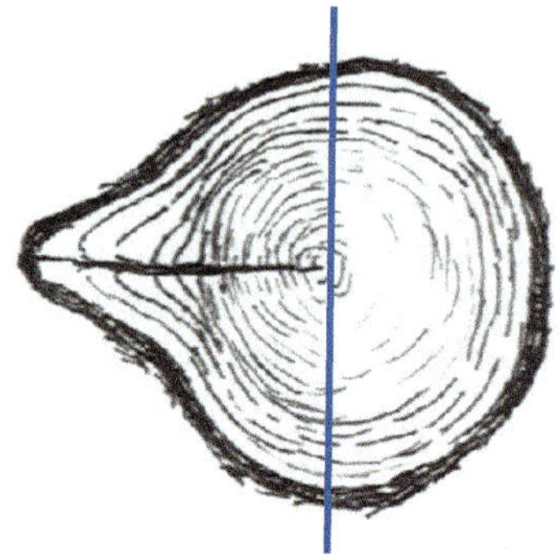

Fig. 7.19 The stem is cut at right angles to the crack. The right part is suited for a veneer flitch

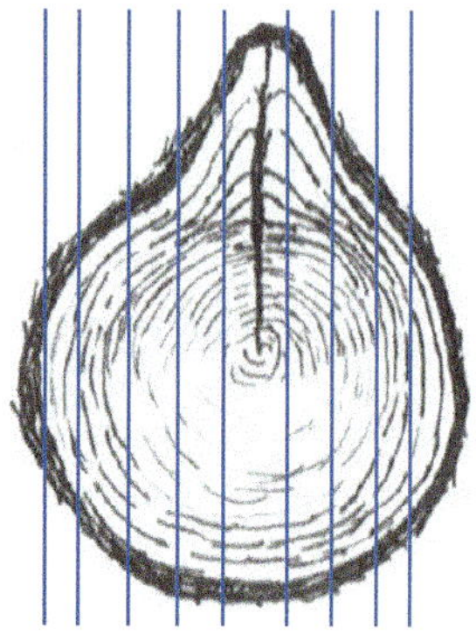

Fig. 7.20 Plain sawn log with frost crack in the heartwood board

7.2 Wind and Snow Effects

7.2.1 Fiber Compressions, Fiber Fracture (Compression Fractures) (Richter 2004a)

7.2.1.1 Description

Fiber compression refers to wood fiber that has been compressed perpendicular to the course of the grain (Fig. 7.21).

Fiber fractures refer to wood fiber that has fractured perpendicular to the course of the grain due to a tree's exposure to mechanical stress which exceeds the compressive strength of the wood (Fig. 7.22). Fiber compressions or fiber fractures typically run in lines or fan out vertically up to an angle of about 60° to the stem axis. They can extend across several tree rings or over as much as 50 % of the stem cross-sectional area (Bues and Stein 1999).

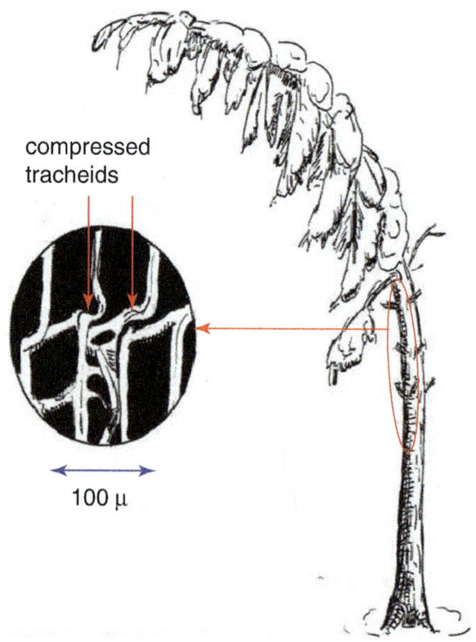

Fig. 7.21 Spruce (*Picea abies*) stem compressed by heavy frost and snow

Each fiber compression is covered on the stem by particularly short-fibered, lignin-rich callus tissue called "*Wulstholz*"

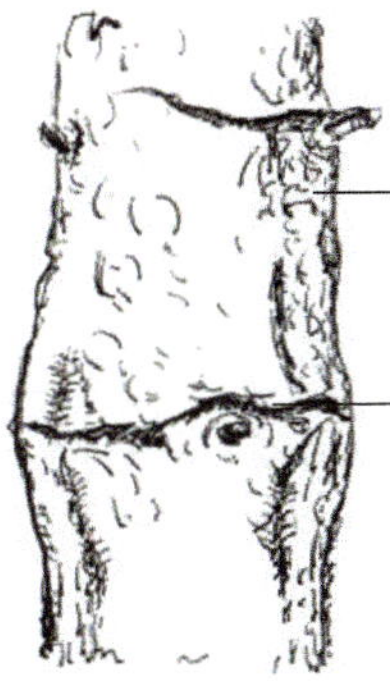

Fig. 7.22 Open fiber fracture in the branch whorl of a spruce (*Picea abies*) with lateral callusing

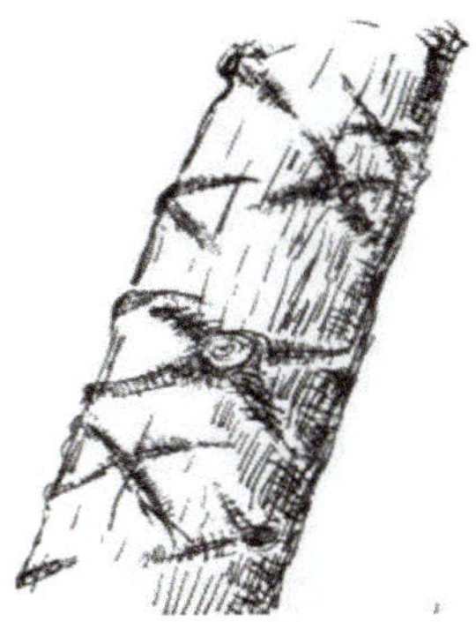

Fig. 7.23 Characteristic "Wulstholz" covering a fiber compression on a debarked spruce stem from an exposed site at high elevation

(bulge wood) (Fig. 7.23). The cells of this "Wulstholz" have cell walls with microfibrils that tend to lean heavily in the longitudinal direction (30°… 50°). This makes the "Wulstholz" more flexible (Rosenthal 2009). Larger fiber fractures can only callus over laterally because the wood at the fracture is no longer vital. It takes considerable skill to be able to distinguish "Wulstholz" from branch scars.

Trendelenburg (1940, 1941) found fiber compressions in 35 species of trees, of which 26 were from tropical and subtropical regions. In Germany, fiber compressions have been detected among others in fir (*Abies alba*), spruce (*Picea abies*), Serbian spruce (*Picea omorika*), larch (*Larix decidua*), and pine (*Pinus sylvestris*) (Helling 1998).

Stein (1998) noted that in spruce (*Picea abies*), two thirds of all fiber compressions occur in the lower third of the stem. The fiber compressions usually begins near branch whorls which are naturally the weakest point in the fiber.

Fiber compressions and fractures can be identified several years after the wood failure by the formation of "Wulstholz" on the compressed side of the stem. As the tree grows in diameter, the callus area disappears, so that eventually trees with fiber compressions and fractures can only be indirectly diagnosed. The likelihood of fiber compressions is high when trees are left growing in stands heavily damaged by wind and snow pressure or isolated on windy sites.

7.2.1.2 Causes

Extreme wind, snow, and ice pressure can force a tree crown to grow towards one side causing the entire stem to bend over. Because the tensile strength of the wood fibers is about twice as large as the compressive strength, an extreme bending stress can cause *fiber compressions* extending beyond the compressive strength zone (Delorme 1974). The stem is compressed even though it is already under a preload for which the compression stress first has to compensate (Kübler 1959). This principle is illustrated, for example, in concrete construction which preloads steel reinforcements.

If the tree is made to swing back and forth by constantly changing pressure loads (gusts), the fiber bands are pulled and pushed until they finally split apart (*fiber fracture*). A compression fracture becomes visible on the stem surface.

Initial starting points for fiber compressions and fractures are the branch whorls. The irregular fiber course found at these spots in the tree presents physically weak areas in the wood (Trendelenburg 1940).

The cambium reacts to the stress by increasing cell growth associated with very wide "Wulstholz" tree rings. Because the "Wulstholz" tracheids are significantly shorter and richer in lignin, they have greater stability. Despite having a higher density than normal wood, "Wulstholz" has a lower modulus of elasticity and compressive strength. "Wulstholz" can be considerably deformed without fracturing but can only withstand limited maximum pressure (Koch 1999).

In the high-density tropical rain forests, intense crown competition forces trees to grow rapidly in the height. This occurs at the expense of radial growth. The height-diameter (h–d) ratios are often over 100:1. Therefore, strong winds can exceed the tree's bending limits resulting in compression fractures (Trendelenburg (1940, 1941).

See Plate 7.4 for causes of fiber compressions and fractures in trees photos and Plate 7.5 for trees with fiber compressions and fractures photos.

7.2.1.3 Prevention

Aside from damaging events such as extremely wet snow conditions or hurricane gusts, fiber compressions are also commonly caused by forest management practices. Taking the following measures can reduce the risk of compression crack damage:

In regions prone to wind and snow breakage, individual trees should, from the early stages of stand development (BHD < 19 cm), be given enough growing space to develop a sufficiently broad crown, connected with a height diameter ratio of under 80:1, which can withstand the extreme bending stresses (Burschel and Huss 1997). However, as the h-d ratio decreases the amount of taper increases.

In addition to the stability of the individual tree, attention must be given to the collective stability of the stand. Thinning

A hurricane with top winds of 125 km/h "selected" all shallow-rooted spruce (*Picea abies*) on pseudogley soil. All trees are suspected to have compression fractures, even the pines (*Pinus sylvestris*) left standing (Tharandt, Germany)

A tornado destroyed this pine-spruce stand (*Pinus-Picea*) within minutes. The trees should not be used as lumber but instead only processed as industrial wood (Quirla, Germany)

Compression fractures are very likely in this mixed stand heavily damaged by frost and snow. Therefore, the wood should not be used as construction lumber (Upper Erz Mountains, Germany)

Storm damaged Monterey pine (*Pinus radiata*) forest (Teneriffa, Spain)

A tornado ravaged a mangrove forest (*Rhizophoraceae*) near the shore of the Cottica River (Commewijne, Surinam)

A storm knocked over this tree in a tropical rainforest. The large height diameter ratio causes significant bending stress and can lead to compression fractures (Surinam)

Plate 7.4 Causes of fiber compressions and fractures in trees

Long-fiber stem fractures on spruce caused by a sudden violent impact (hurricane) (*left*) are very different from a flat fiber break caused by frequently changing pressure loads (*right*) (Photo on the *right*: G.-E. Weber)

One growing period after a hurricane. Spruce (*Picea abies*) with visible fiber fracture (*left*). Tree in the same stand with "Wulstholz" (*right*) (*Germany*) (Photo: G.-E. Weber)

Fiber compressions and fractures incured during a hurricane ultimately break the stem of this linden (*Tilia* ssp) (Germany)

Despite its wide-spreading buttress roots, this narrow-stemmed zwart riemhout stem (*Micropholis guyanensis*) suffered a fiber fracture – evidence of severe bending stress (Surinam)

Heavy "Wulstholz" formation on a debarked spruce (*Picea abies*) on an exposed, high elevation site (800 m) (Upper Harz, Germany)

Pine (*Pinus sylvestris*) with "Wulstholz" (Muddus, Jokkmokk, Sweden)

Split spruce (*Picea abies*) stem with compression folds and "Wulstholz"

Plate 7.5 Trees with fiber compressions and fractures

operations, especially in older stands, should not overly interrupt the crown closure.

The most stabile stands have a mix of species and a vertical structure (Fig. 7.24).

The spatial arrangement of the trees must be carefully managed so that the stand's protective cover remains intact.

7.2.1.4 Impact on Use

Softwood stems, particularly spruce (*Picea abies*) on the edges of snow and wind damage stands or sites with extreme exposures, are considered particularly vulnerable to fiber compressions.

Fiber compressions that are already visible on the outer bark of the stem and compression fractures significantly affect the strength of the wood (Fig. 7.25). Sawn timber with such damages should not be used for constructive purposes or for load-bearing functions (EMPA 1990) (Fig. 7.26). The same applies to smaller pieces of wood used, for example, for roof battens or telephone poles. It is not uncommon for impregnated telephone poles, which may at first sight appear perfectly sound, to split apart during installation. The injured spot was obtuse and shell shaped – a sure sign of an extensive fiber compression that went unnoticed during the initial stem diagnosis.

Therefore, if callusing on the stem surface reveals fiber compressions and fractures in the wood, it is important that these factors be taken into account during grading. According to the general quality standards stated in the HKS, wood with such defects cannot be classified as in A/EEC or B/EEC grade, but can still be used as C grade wood (DIN 2002b, p. 12). Nevertheless, because it is possible that wood could accidentally be use for construction purposes, it is best classified as industrial wood (e.g., pulpwood).

Following CEN standards, such logs are only classified as D grade wood since the fiber fracture "…impairs the natural wood properties…" of a significant portion of the stem cross-sectional area (DIN 1999b, p.4).

7.2.1.5 Technological Adaptation

Under visual diagnosis, wood with fiber compressions which are only visible on planed wood are considered safe to use. These fiber compressions occur at a bending stress of about 50 % of the breaking load. However, macroscopically visible fiber compressions can also occur at a bending stress of about 75 % of the breaking load (Kisser and Steininger 1952). In this case, the wood should no longer be used for static purposes.

It is possible using strength measuring instruments to easily detect compression fractures in lumber as they usually appear as the "weakest section." The elastic modulus of the wood is calculated from the test set, dimensions, proof load, and deflection of the lumber. This correlates with the bending strength, which is a measure of the strength of the sawn timber (Fig. 7.27) (DIN 2001a).

Ultrasound tests for detecting compression fractures in squared timber are still inconclusive (Kessel 1990).

See Plate 7.6 for identifying fiber compressions and fractures in processed wood photos.

Fig. 7.24 Vertically structured mixed stands, combined with suitable forest boarder plantings, are better protected from the mechanical stress of extreme weather events than single-age stands

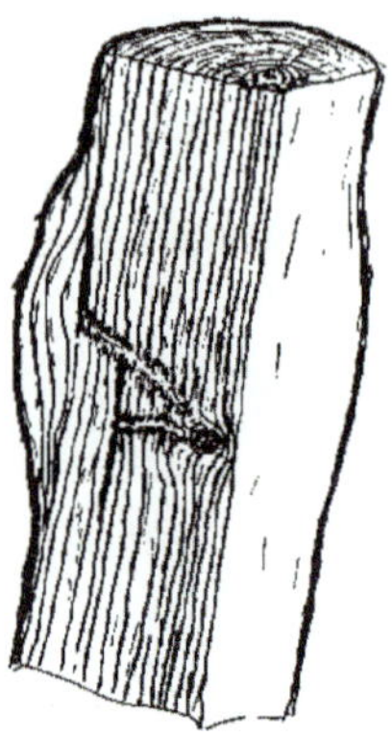

Fig. 7.25 Log with compression folds and "Wulstholz"

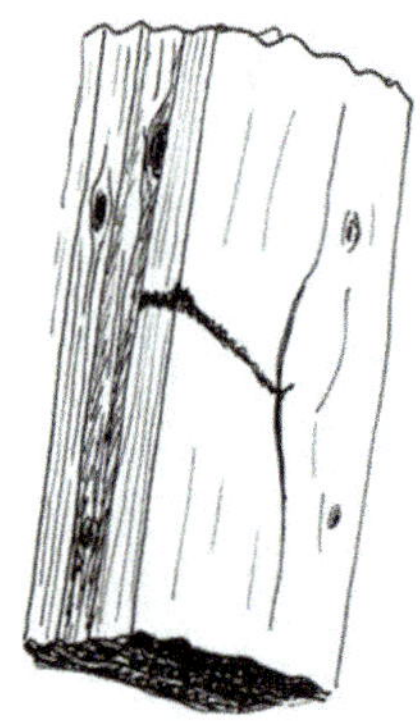

Fig. 7.26 Fiber fractures in a piece of composite lumber

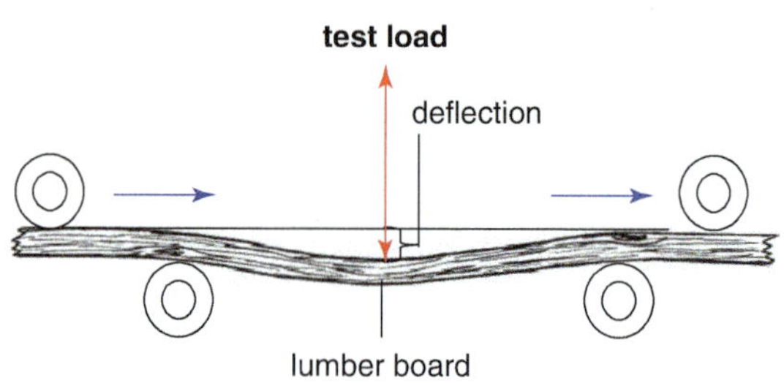

Fig. 7.27 Principle of stress grading

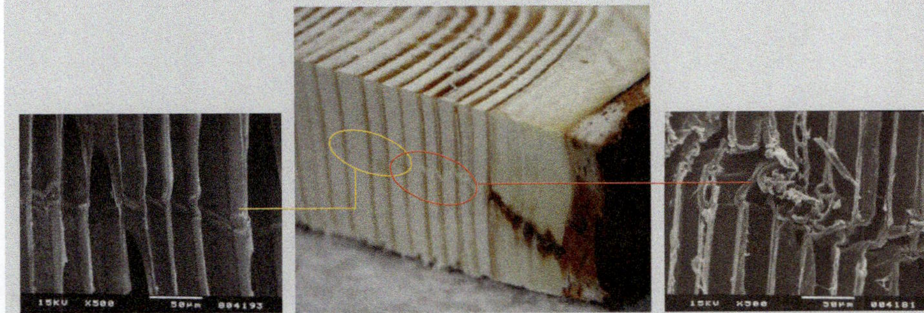

Electron microscopic images of mild compressions to the cell walls of spruce tracheids (*left*) and pressed tracheids (*right*) (Photos: E. Bäucker). In the middle image the position of the compressions to the spruce fibers can be roughly determined based on the compression folds and "Wulstholz" (Photo: G.-E. Weber)

The fiber compressions in this spruce (*Picea abies*) board only become visible after planning. It can still be used for static purposes

Fiber fracture, the spruce (*Picea*abies) board is no longer useable for static purposes

Fiber fractures revealed during processing on this wanakwari board (*Vochysia tomentosa*) (Surinam)

The wooden posts of this new sawmill made of kwatapatu (*Lecythis zabucajo*) (links) show partially of fiber compression (*center*) or fiber fractures (*right*). While the fiber compressions do not impact the statics of the building, such extensive fiber fractures are problematic (Surinam)

Fiber compressions (*left*) and fiber fractures (*right*) on a post made of walaba (*Eperua* ssp.). The fiber fractures significantly reduce the residual strength of the pillar's cross section (Surinam)

Plate 7.6 Identifying fiber compressions and fractures in processed wood

7.2.2 Shear Stress Cracks (Richter 2006c)

7.2.2.1 Description

Shear stress cracks appear as cross-sectional shakes in older softwoods, especially in spruce (*Picea abies*). They can extend several meters through the lower part of the stem (Fig. 7.28). This form of shake typically runs across the entire stem. The first indicator of such damage is resin discharge along the course of the cracks on both sides of the stem surface.

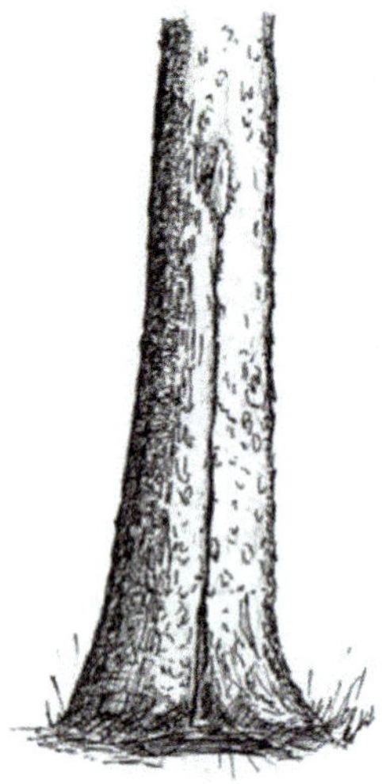

Fig. 7.28 Spruce (*Picea abies*) with resin discharge caused by shear stress cracking

In larch (*Larix decidua*), shear stress cracks appear either as resin-filled heart shakes at the stem base (pitch pockets), which have not (yet) reached the sapwood, or as a split stem (Fig. 7.29).

Especially in pines (*Pinus* ssp.), but also in fir (*Abies alba*), spruce (*Picea abies*), and Douglas fir (*Pseudotsuga menziesii*), shear stress cracks appear as a heart shake near the center of the butt log. These cracks which run from the radius of the stem are referred to as schilfer cracks.

All the cracks forms mentioned above originate in standing trees.

See Plate 7.7 for shear stress cracks photos.

Fig. 7.29 Larch (*Larix decidua*) with sweep and split stem

130 year old spruce (*Picea abies*) forest with beach (*Fagus sylvatica*) understory; a clear cutting had originally been planed 15 years earlier. This stand is at risk of shear stress cracking (Germany)

Shear stress leads to an initial crack with resin discharge in this spruce (*Picea abies*)

In the course of a few years, visible shear stress cracks form on the opposite side of this wind exposed spruce (*Picea abies*) stem

Isolated spruce (*Picea abies*) with shear stress cracks on an exposed site at a high elevation (Thuringian Forest, Germany)

High reaching, double-sided shear stress crack on an old spruce (*Picea*) on a rocky site (Tharandt, Germany)

Spruce (*Picea abies*) with lopsided crown on a slope with shear stress crack (Tharandt, Germany)

Broken spruce (*Picea abies*) log split along the former shear stress crack (Tharandt, Germany)

Plate 7.7 Shear stress cracks

Predominate larch (*Larix decidua*) with stem sweep and open shear stress crack

Larch (Larix) with heart crack formed into pitch pockets is likely due to shear stress

Open shear stress crack on a tilted Douglas fir (*Pseudotsuga menziesii*) with lopsided crown at 2300 m (Wyoming, USA)

Spruce (*Picea abies*) board with stepped schilfer shake caused by shear stress in the stem

Oak (*Quercus* ssp.) with the so-called "bad luck beams" The shear stress split this heavily cantilevered branch (Tharandt, Germany)

Bad luck beam in a beech (*Fagus sylvatica*). In the winter snow and frost loaded on the branch. Shear stresses split the branch. East Carpathians, 1,400 m elevation (South Western Ukraine)

Plate 7.7 (continued)

7.2.2.2 Causes

Shear stress cracks can be caused by wind pressure resulting in one-sided crown formation and tilted stems particularly in dominant trees and in old softwoods (Fig. 7.30). Constant unilateral stress and crown sway causes shear stress, particularly at the base of the stem. This stretches the concave buttress roots on the tension side of the stem. If the transverse tensile strength of the tree's fiber network is exceeded, the stem will crack (Fig. 7.31).

A further condition for shear stress cracks is a solid ground surface on the compressed side of the roots (shallow soils with solid "zersatz" zone, rocky subsoil). Roots on the tension side must be well anchored in the ground. On the one hand, the stable anchoring prevents windfall. On the other hand, the shear pressure is reduced so that it is no longer strong enough to break the stem.

Bole sweep also causes shear stress cracks, especially if the main amount of pressure is directed towards the concave side of the stem. This applies particularly for certain larch (*Larix decidua*) provenances prone to sweep. If the crack is only marked by a heart shake, it fills up with resin that emerges from the split resin canals in the cavity and forms pitch pockets.

Schilfer cracks are heart shakes in old softwoods, especially pine (*Pinus sylvestris*) and fir (*Abies alba*), and are a result of bending stress. In this case, the dry heartwood cracks due to insufficient elasticity. The cracks are not outwardly visible (Volkert 1940). They show up in lumber when it is cut from a spiral grained stem.

The reasons given above still do not sufficiently explain all the causes for schilfer cracks and pitch pockets.

Far reaching, cantilevered branches can also form shear stress cracks. They result from stress caused by the branches' own weight, in combination with snow cover and wind. These longitudinally cracked branches are also known as "bad luck beams." They appear more frequently in deciduous than in coniferous trees.

7.2.2.3 Prevention

When logging and clearing activities cease in single-age stands, as often the case with stand conversions, it commonly leads to over mature pure softwood stands. These old trees should be removed immediately since their large crowns on the former felling face are stressed by wind causing new shear stress cracking.

Trees towering above the canopy, or trees suddenly exposed, are also vulnerable to shear stress. Therefore, it is important to ensure adequate coverage within the stands and between the stands.

Slender branched trees are often promoted for wood quality purposes. However, under adverse site conditions, this can also lead to the formation of shear stress cracks because there is no "anchor effect" from the stronger branches in the lower sections of the stem.

Larch provenances prone to bole sweep are not suitable for new stand plantations.

Thinning measures must be carried out in way that ensures even crown development.

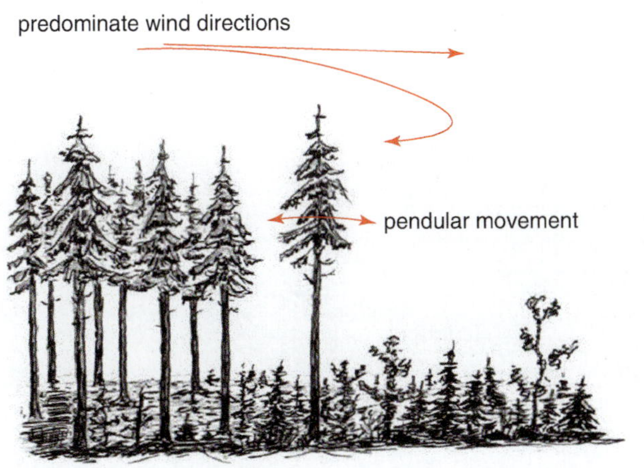

Fig. 7.30 Over mature spruce (*Picea abies*) on an inactive clearing site should be harvested immediately

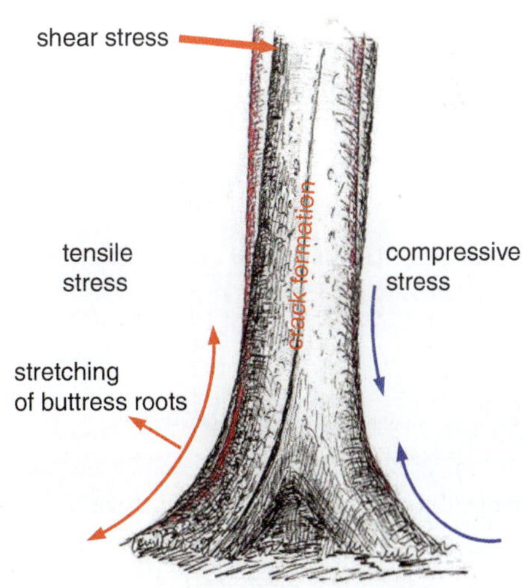

Fig. 7.31 Principle of shear stress crack formation

7.2.2.4 Impact on Use

Transversal cracks in spruce (*Picea abies*), larch (*Larix decidua*), Douglas fir (*Pseudotsuga menziesii*), and schilfer cracks in fir (*Abies alba*), spruce (*Picea abies*), and pine (*Pinus sylvestris*) devalue the most valuable part of the stem base, because they usually run through the main lumber products (beams, planks, boards) (Fig. 7.32 and 7.33).

After the resin canals running out from the radius are broken, the resin flows into the cracks (pitch pockets in larch (*Larix)* (Fig. 7.34). Any wood processing is complicated by resin buildup on the equipment. The affected area of timber prevents any higher use.

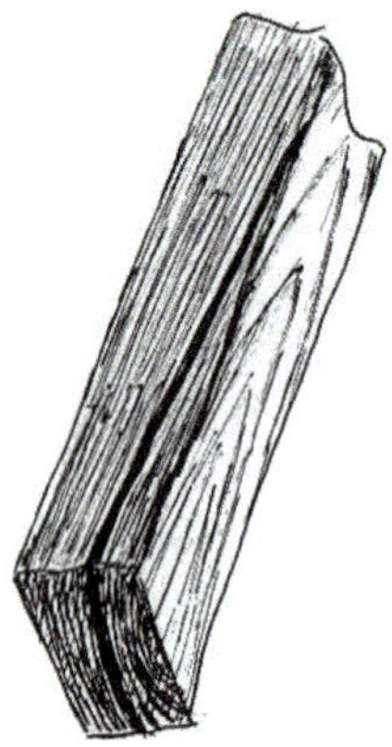

Fig. 7.32 Board with shear stress crack

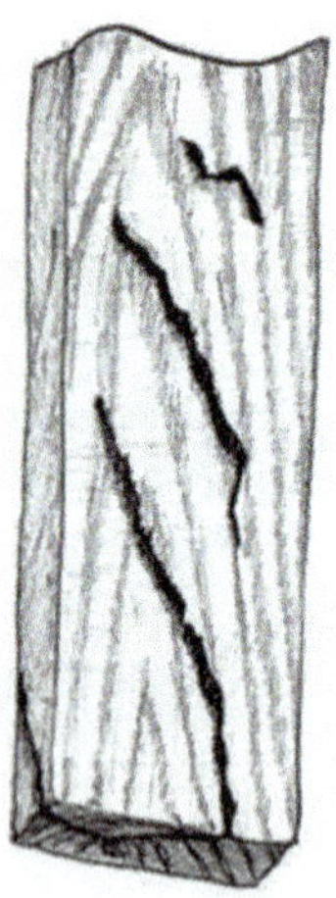

Fig. 7.33 Fir (*Abies alba*) board with schilfer shake

Schilfer cracks are often still hidden in the lumber and only revealed during finishing (planting, milling). Defects in the finishing significantly affect the operating profit.

7.2.2.5 Technological Adaptation

Stem wood with cross-sectional cracks should be scaled so that as much of the crack plane as possible lies in the same direction as the cutting face and only in one board or plank (Fig. 7.35).

Fig. 7.34 Larch (*Larix decidua*) stem with split pitch pockets

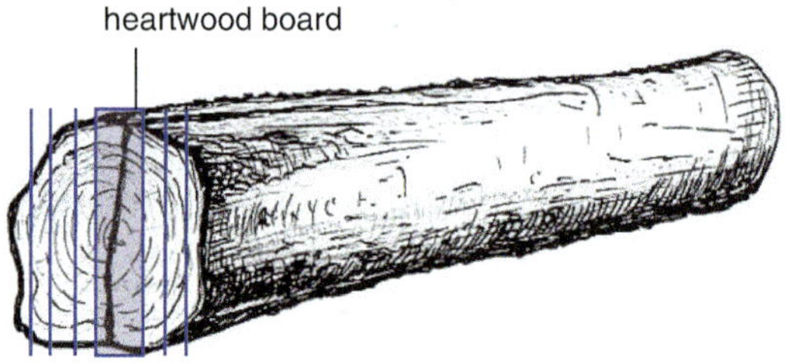

Fig. 7.35 Plain sawn log with shear stress crack

8 Overview of Cracks/Shake Forms and Causes

See Plate 8.1 for didactic reasons Table 8.1 overview of cracks forms and causes photos.

Knuchel (1934), Mette (1984), Richter (2010), Steuer (1990), and Wagenführ and Scheiber (1989).

C. Richter, *Wood Characteristics: Description, Causes, Prevention, Impact on Use and Technological Adaptation*,
DOI 10.1007/978-3-319-07422-1_8,

Table 8.1 Overview of cracks/shake forms and causes

Description of shakes based on form	Description of shakes based on cause	Illustration of shakes
Heart shake/pith shake: appears as a *single shake*, *T-shake*, *Y-shake*, *cross shake*, or *star shake* running through the pith Crack extending radially from the pith, but not to the stem surface In old spruces (*Picea abies*), larches (*Larix decidua*), pines (*Pinus sylvestris*), and firs (*Abies alba*), heart shake can extend several meters up from the stem base Trees from tropical rain forests often crack after felling due heavy growth stress. Heart shake usually only occurs during the short transition to the cross-sectional surface crack	1. *Dry crack*: occurs when wood dries out below the fiber saturation point of approximately 30 %. This can happen in heartwood or sapwood shortly after felling in the drier core of the stem (initial cracking) 2. *Schilfer shake:* occurs mainly in mature softwoods. Wind and snow pressure bends the trees so severely that tangential cracks form in the drier core area of the living tree. These cracks are only visible in sawn timber 3. *Pitch pockets*: shear stress (wind pressure) applied to living larch (*Larix decidua*) trees with bole sweep can cause shakes to form on the concave side of the stem that eventually fill up with pitch (see Sect. 7.2.2)	**Fig. 8.1** Cross shake appearing as heart shake in oak (*Quercus* ssp.)  **Fig. 8.2** Schilfer shake in a fir (*Abies alba*) board 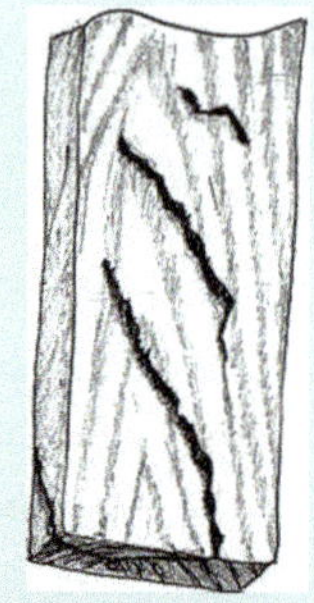 **Fig. 8.3** Heart shake, appearing as pitch pocket in larch (*Larix decidua*) with eccentric pith 
Traversing shake: appears as a single shake, T-shake, Y-shake, cross shake, or star shake running through the pith Crack in a felled tree extending from the pith and reaching the stem surface. Particularly common in beech (*Fagus sylvatica*), poplar (*Populus* ssp.), and willow (*Salix* ssp.)	1. *Dry crack*: occurs when the wood dries out below the fiber saturation point of approximately 30 %. Originates at the pith and extends to the stem surface (see seasoning crack illustration) 2. *Stress crack*: growth stress cracks which rupture after a felling cut is made to the stem base. They emerge from the swelling of the lignin and cellulose contraction in the cells of the youngest tree ring or increment zones. This leads to such a degree of tension stress in the stem surface and compression stress in the core that the wood cracks shortly after felling or during processing (see Sect 3.3) Note: stress cracks are not seasoning cracks!	**Fig. 8.4** Stress crack appearing as traversing shake in beech (*Fagus sylvatica*) 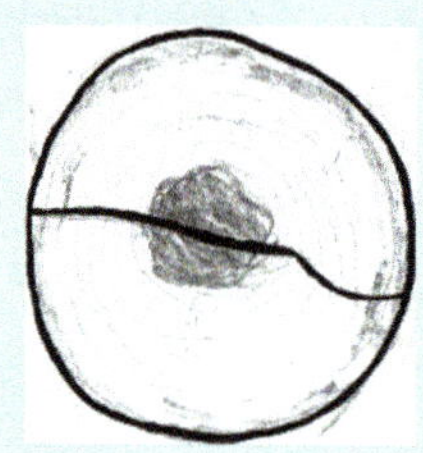 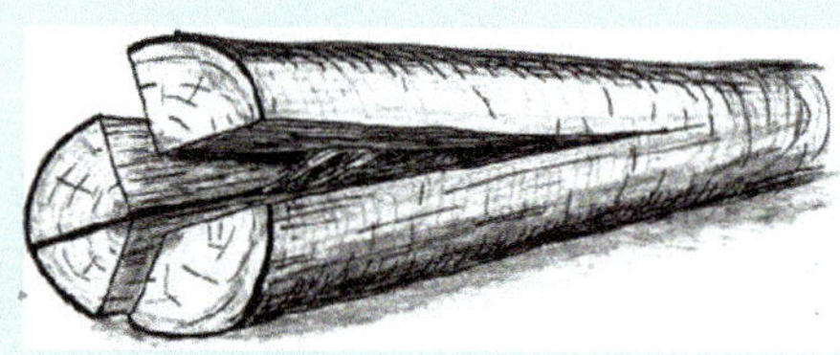 **Fig. 8.5** Stress cracks appearing as cross shake in beech (Fagus sylvatica)
Seasoning crack: appears as a *combination* of a *single shake, T-shake, Y-shake, cross shake, star shake, radial shake*, and *ring shake* Thin radial cracks extending across the entire cross-sectional area of a felled tree stem or stem section that begin to penetrate deeper into the wood as it dries	*Dry crack:* rapid drying of freshly cut, debarked wood (summer harvest) and dead wood far below the fiber saturation point of approximately 30 %. The anisotropy of wood causes it to shrink during drying longitudinal-radial-tangential in the ratio of 1:10:20. This leads to radial shakes. Wood rays act as additional fissure lines	**Fig. 8.6** Dry cracks appearing as a combination of radial shakes, single, radial running cracks, a Y-heart shake, and a ring shake in ash (*Fraxinus excelsior*) 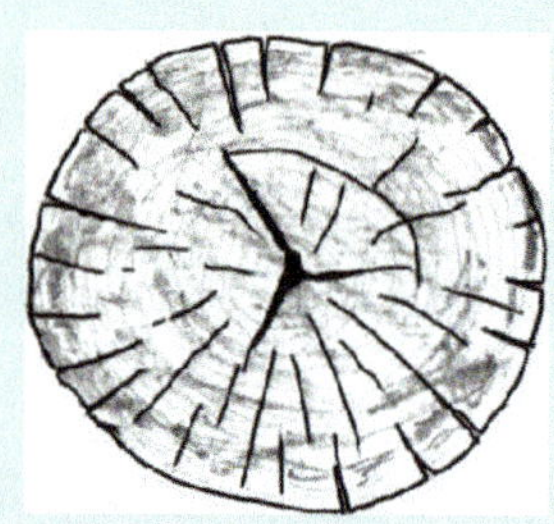

Table 8.1 continued

Description of shakes based on form	Description of shakes based on cause	Illustration of shakes
Ring shake/cup shake A crack in the wood parallel the tree rings, especially between the rings that contrast significantly in size. Common in suddenly exposed very old spruce (*Picea*), fir (*Abies*), oak (*Quercus*), poplar (*Populus*) and willow (*Salix*). In tropical trees, ring cracks typically occur during the transition from the juvenile to the mature increment zones	*Stress crack:* varied degrees of swelling and shrinking with irregular tree ring formations and extreme differences in density (open stand increment, reaction wood) also in the boundary between juvenile and mature wood. This can cause a ring shake to form while a tree is still living. The shake only becomes visible after the tree is felled on the crosscut surface. Ring shake has *physical* causes	**Fig. 8.7** Ring shake in a section with extremely varied tree ring widths in larch (*Larix decidua*) 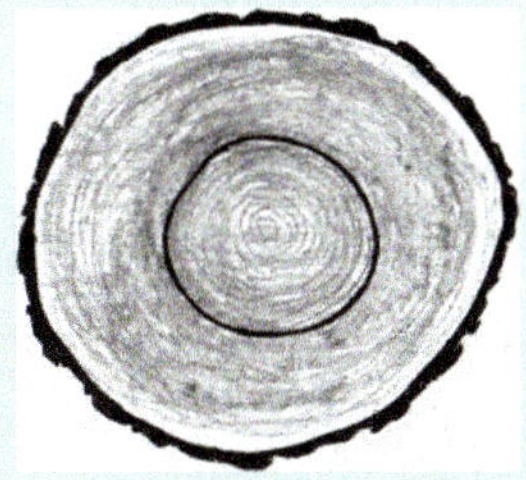
Ring rot Pathological cracking parallel to the tree rings, particularly between the wide rings in oak (*Quercus*) and chestnut (*Castanea sativa*). One or several of these rings are structurally damaged by fungal infestation	*Ring shake with fungal infestation/ defective core:* the large early wood vessels of the individual, insufficiently formed tree rings (early frost) of ring-porous wood is infested by damaging fungi (*Phellinus*). In ring shake the xylem encircled by the shake completely separates from the rest of the stem along the damaged tree rings. Ring failure has *pathological* causes	**Fig. 8.8** Oak (*Quercus* ssp.) with ring shake caused by fungal infection 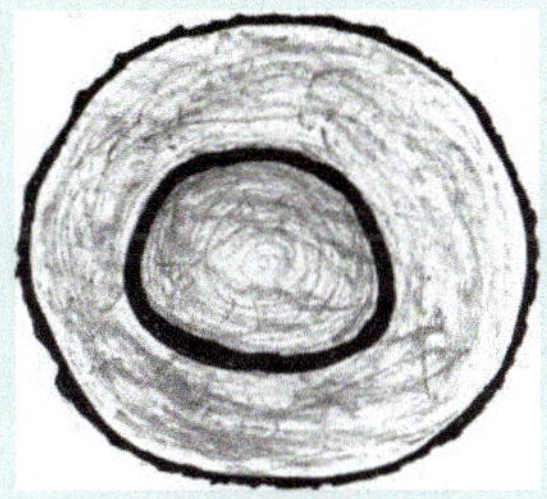
Loose heart Stems with severe heartwood decay caused by microorganism infestation	*Ring shake with decayed heartwood:* common especially in tropical tree species of the primary rain forest. Young trees try to use as little material as possible to outgrow light competitors by not storing any heartwood substances in their juvenile wood. The heartwood only becomes durable when it is mature. The juvenile core becomes loose (physically) as a result of the changes in anatomical structure during the long period to maturity and is destroyed by microorganisms (pathological)	**Fig. 8.9** Loose heart on kopi (*Goupia glabra*) 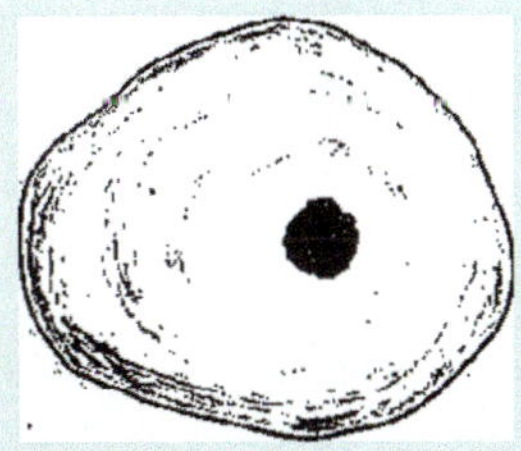  **Fig. 8.10** Longitudinal cut on kopi (*Goupia glabra*) with loose heart
Spider shake: appears as a combination of *radial running star shakes and ring shakes* parallel to the tree rings or increment zones in tropical tree species	*Dry crack:* dehydration of adjacent tree rings with varied widths (ring shake) and the heartwood zone (radial shake). The crack originates in standing trees and become visible when the fiber saturation point drops below approximately 30 % after a long storage period (see also "seasoning crack" and "ring shake")	**Fig. 8.11** Oak with spider shake caused by uneven drying process 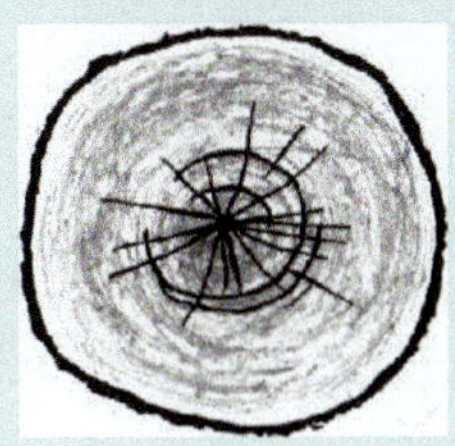

Table 8.1 continued

Description of shakes based on form	Description of shakes based on cause	Illustration of shakes
Coat shake Crack running with the course of the fiber on the stem surface. Can originate in standing trees	1. *Frost crack:* occurs when the temperature suddenly falls below the freezing point. The shock-like cooling leads to nonuniform thermal contraction. The radial-tangential anisotropy of wood, as well as the effect of the freeze-drying process increases the stress. The crack can extend to the pith (see Sect. 7.1.3)	**Fig. 8.12** Cross section from a maple (*Acer* ssp.) stem with frost scar and frost crack extending to the pith 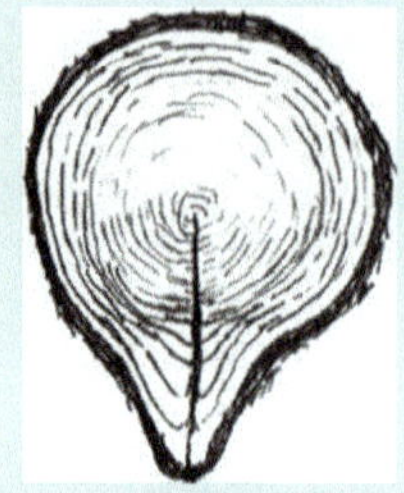
	2. *Heat crack:* results from drought stress in living trees (flat-rooted spruces) during periods of very hot and dry weather. This can decrease the moisture content of sapwood to below 30 % 3. *Dry crack:* occurs when dead wood dries to below the fiber saturation point of approximately 30 %. As the crack forms, wood rays act as fissure lines (see also "seasoning crack")	**Fig. 8.13** Maple (*Acer* ssp.) with repeatedly ruptured frost crack and frost scar 
Tangential shake: appears as *cross-sectional cracks* that do not extend to the pith Cross-sectional crack starting in the stem base and extending several meters up the stem	*Shear stress cracks:* wind and snow pressure can force the tree to bend so far over that a tangential running crack forms (see Sect. 7.2.2) Growth stress in tropical species can cause the stem to crack in segments. This also leads to tangential shake	**Fig.8.14** Tangential shear stress crack in a spruce (*Picea abies*) (top) and larch (*Larix decidua*) stem (left) 

Table 8.1 continued

Description of shakes based on form	Description of shakes based on cause	Illustration of shakes
Fiber crack: appears as cross-sectional shake Fiber cracks are ruptures in the wood fiber running with the course of the fiber in standing trees, often callused over with "Wulstholz." Particularly common in spruce, less in other softwoods: also occur in tropical tree species	*Fiber fractures:* trees sway back and forth under alternating degrees of stress, pressing and pulling the wood fibers until they ultimately break. The resulting compression fracture is visible on the stem surface (see Sect. 7.2.1) If tropical trees are exposed to bending, their usually large height-diameter ratio leads to fiber fractures	**Fig. 8.15** Spruce (*Picea abies*) with fiber fractures. Viewed externally 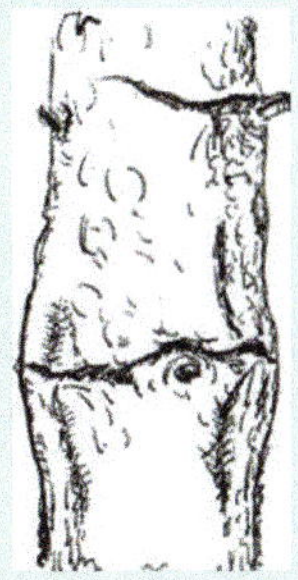 **Fig. 8.16** Fiber fractures in a spruce (*Picea abies*) radial section 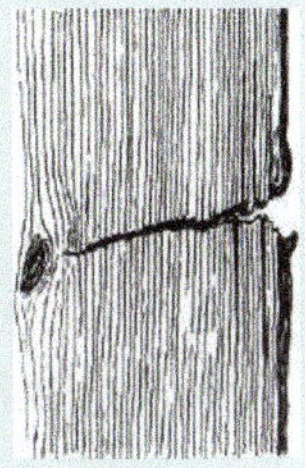
Radial shake: normally one tree ring wide Spindle-shape crack running radially within a tree ring	*Suction cracks:* caused by suction tension brought on by spring transpiration and impeded water absorption due to aridity or ground frost (drought stress), especially in immature trees with wide tree rings and tree rings with low density (see also "pitch pockets")	**Fig. 8.17** Spruce (*Picea abies*) with suction cracks 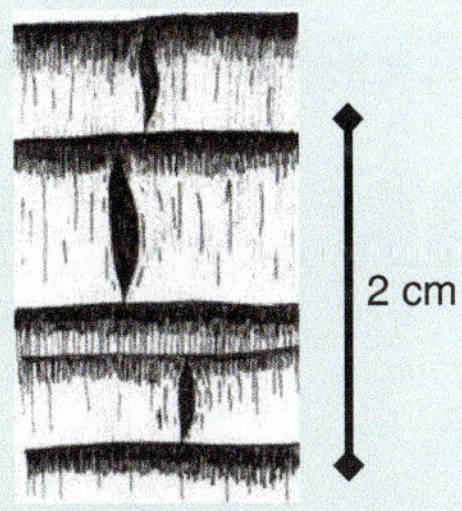
Ring shake within a confined section of a tree ring Resin-filled cavities within the tree rings of species with resin canals (spruce (*Picea*), larch (*Larix*), Douglas fir (*Pseudotsuga*), and pines (*Pinus* ssp.) In tropical trees in addition to resins cracks can be filled with gums, kinos, latices	*Pitch pockets/pitch shake:* caused by a hydraulic effect triggered by defects in the resin canals, particularly in wide lumen earlywood rings where the freed resin presses between two adjacent tree rings separating them from each other. This defect is attributed to increased suction tension (see also "suction cracks") Why in some tropical species exudates flow between cracks and increment zones is not fully clear, but likely linked the closure of internal injuries	**Fig. 8.18** Spruce (*Picea abies*) with pitch pockets 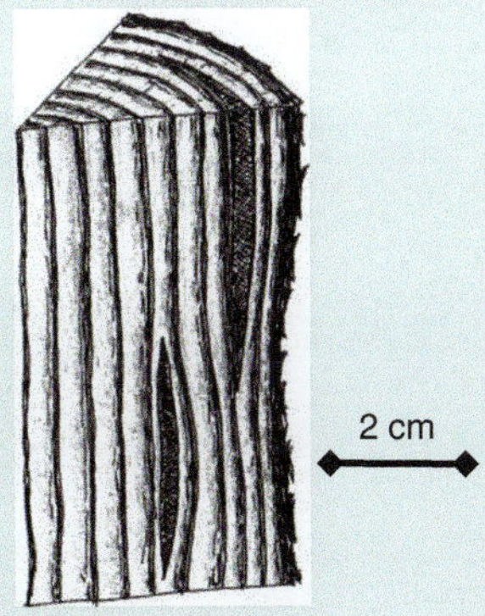

Veneer oak (*Quercus petraea*) logs with new heart shake (initial crack) caused by letting the heartwood dry out below the fiber saturation point (Saxony, Germany)

Gele kabbes (*Fava amargosa*) with cross shake in the pith (initial crack) (Surinam)

Larch (*Larix decidua*) log with heart shake (pitch pocket) due to shear stress (?)

Schilfer shake as tangential crack inside a spruce (*Picea abies*) beam

Cross-sectional crack in an oak (*Quercus petraea*) caused by growth stress. A metal band prevents further cracking (Lower Saxony, Germany)

Traversing shake caused by growth stress on a maple (*Acer platanoides*) log with plastic brackets to prevent further cracking (Saxony, Germany)

Y-shake caused by growth stress in a beech (*Fagus sylvatica*) log with metal brackets (Saxony, Germany)

Newly cut oak (*Quercus petraea*). The T-shake formed as a result of growth stress; if left unbraced, further cracking is likely (Saxony, Germany)

Fresh cross shake caused by growth stress on a beech (*Fagus sylvatica*) veneer log. Plastic brackets failed (Saxony, Germany)

Plates 8.1 Crack forms and causes

Beech (*Fagus sylvatica*) log with star shake caused by growth stress

Oak (*Quercus* ssp.) with taping cross-sectional crack due to growth stress (tension wood) (Baden-Württemberg, Germany)

Growth stress tangentially cracked this donsedre stem (*Cedrelinga cateniformis*) (Surinam)

These wanakwari *(Vochysia tomentosa) boards craked under growth stress at the mill* (Surinam)

Fluted birch (*Betula pendula*) with seasonal cracks (star shake) due to excessive drying

Right spiral pine (*Pinus sylvestris*) stem with radial shake caused by a dry crack (Muddus, Sweden)

Left spiral spruce (*Picea abies*) with a dry crack extending to the pith

Larch (*Larix*) with ring shake at the transition from tree rings of varied widths, visible as log begins to excessively dry out

Ring failure (ring shake with fungal infestation) in a veneer oak (*Quercus* ssp.) log caused by incomplete heartwood formation in adjacent tree rings

Plates 8.1 (continued)

Veneer oak (*Quercus petraea*) with spider shake visible as log start to dry out (Saxony, Germany)

Ash (*Fraxinus excelsior*) log with seasonal cracks partly (spider shake) due to excessive drying (Tharandt, Germany)

Ring shake in a freshly cut basralocus (*Dicorynia guianensis*) log (Surinam)

Radial shakes in a basralocus (*Dicorynia guianensis*) log (*left*) and pakuli (*Platonia insignis*) log (*right*) form after several years of storage. The ring shake marks the transition from juvenile to mature wood (Surinam)

Heart rot is already visible in the juvenile wood of this bruinhart stem (*Vouacapoua americana*) (Surinam)

These kopi (*Goupia glabra*) (*left*), bolletrie (*Manilkara bidentata*) (*center*), and greenheart (*Tabebuia serratifolia*) stems come from dense primary forest stands. The heart rot is clearly visible in the juvenile rapidly grown wood with minor heartwood formation, while the mature wood in the upper stem sections is fully developed (Surinam)

Plates 8.1 (continued)

Oak (*Quercus robur*) stem with coat shake caused by a fresh frost crack

Frost cracks in a maple (*Acer pseudoplatanus*)

A period of drought led to a heat crack (radial shake) in this shallow rooted spruce (*Picea abies*)

Larch (*Larix decidua*) with traversing shake caused by a shear stress crack

Tangential cracks in a wiswiskwari (*Vochysia guianensis*) log due to growth stress (Surinam)

Jongo kabbes (*Vataireopsis speciosa*) with tangential and radial craks. The log is cut at the level of the radial crack (Surinam)

Spruce (*Picea abies*) log with pitch pocket between two tree rings leading to "ring" shake . Radial section

Spruce (*Picea abies*) board with tangentially arranged pitch pockets

Ring and radial shake in a rodelokus (*Hymenaea courbaril*), filled with resin (Surinam)

Fatty exudate flows out of the cross-sectional crack in a walaba stem (*Eperua* ssp.) (Surinam)

Plates 8.1 (continued)

Resin-filled larva holes in the ring shake of a pakuli (*Platonia insignis*) (Surinam)	Fiber fractures in a zwart riemhout (*Micropholis guyanensis*), visible as a surface cross-sectional crack (Surinam)	Radial shake through a fiber fracture in a spruce (*Picea abies*) with "Wulst-holz" (*left*), visible as a surface cross-sectional crack (Erz Mountains, Germany)	Radial section of a fiber compression in a walaba (*Eperua* ssp.), visible as a crack in the surface cross section (Surinam)	Suction crack as a radial shake in a spruce (*Picea abies*) (Germany)

Plates 8.1 (continued)

Erratum to: Wood Characteristics

Description, Causes, Prevention, Impact on Use and Technological Adaptation

Christoph Richter

DOI 10.1007/978-3-319-07422-1

In the original version the University and Faculty of the author were missing in the following sentence:

Then, after high school, he studied in Tharandt from 1963 to 1966.

The correct sentence is:

Then, after high school, he studied at the Dresden University of Technology, Faculty of Forestry in Tharandt from 1963 to 1966.

The online version of the original book can be found under
DOI 10.1007/978-3-319-07422-1

C. Richter, *Wood Characteristics: Description, Causes, Prevention, Impact on Use and Technological Adaptation*,
DOI 10.1007/978-3-319-07422-1_9,

References

Aaftrink H, Hasse P, Weiß A (2005) Leben mit Blitzen. Winterthur-Versicherung, Winterthur

Altenkirch W, Majunke C, Ohnesorge B (2002) Waldschutz auf ökologischer Grundlage. Ulmer, Stuttgart

Anonymus (2000) Quo vadis? – Empfehlungen zum Markieren von Wanderwegen. Dresden, Landestourismusverband Sachsen (publ.)

Anonymus (2003) www.top-wetter.de/lexikon. 10.04.2005

Anonymus (2005) www.netzwelt.de/lexikon. 28.04.2005

Autorenkollektiv (1988) Lexikon der Holztechnik, 3th revised edn. Leipzig, Fachbuchverlag

Autorenkollektiv (2006) Auswirkungen von Rotwildschälschäden auf den finanziellen Ertrag eines Fichtenbestandes. Paper. Lehrstuhl Forstnutzung, Technische Universität Dresden

Baatz H (1985) Mechanismus der Gewitter und Blitze, vol 34. VDE-Schriftenreihe, Berlin, p 17

Bachmann P (1970) Wirtschaftliche Überlegungen zur Waldpflege. HESPA Mitt 1:1–24

Baltrusch M (2003) Ermittlung von Auswahlkriterien für Resonanzholz für Streichinstrumente. Research report. Inst f Musikinstrumentenbau Zwota an der Technischen Universität Dresden

Berger K (1966) In: Aaftrink H, Hasse P, Weiß A (2005) Leben mit Blitzen. Winterthur-Versicherung, Winterthur

Binder B (2008) Hilfstafeln zur Erhebung und Bewertung von Schälschäden an Fichte, basierend auf Untersuchungen der Forstlichen Bundesversuchsanstalt Wien. Diverse Veröffentlichungen/Forstliche Bundesversuchs-anstalt 17:9

Bosshard HH (1984a) Holzkunde Bd 2. Birkhäuser, Basel/Boston/Stuttgart

Bosshard HH (1984b) Aspekte der Holzbearbeitung und Holzverwertung, 2 revised ednth edn. Birkhäuser, Basel/Boston/Stuttgart

Briedermann L (1986) Schwarzwild. Deutscher Landwirtschafts-Verlag, Berlin

Brienen RJW (2013) Jaarringen in de tropen. Vortrag CELOS, Paramaribo 15. März 2013

Brienen RJW, Helle G, Pons TL, Guyot J-L, Gloor M (2012) Oxygen isotopes in tree rings are a good proxy for Amazon precipitation and El Niño-Southern Oscillation variability. www.pnas.org/lookup/suppl/doi:10.1073/pnas.1205977109/-/DCSupplemental. 16.03.2013

Buchner E (2008) Zukunft im Dialog zwischen Theologie und Naturwissenschaft. Kurzreferat im Theologischen Forum, Lehrstuhl f. Genetik, Universität Würzburg

Bues C-T, Stein A (1999) Erkennen und richtig handeln: Faserstauchungen und Faserbrüche im Holz von Fichten. AFZ-Der Wald 15:796–797

Bues C-T, Weiß M (2002) Zur Holzqualität unterständiger Häher-Eichen in sächsischen Kiefernbeständen. In: Proceeding: Waldumbau im globalen Wandel, Forstwiss Tagung Göttingen, p 220–221

Bues C-T, Braun M, Reubel M (1989) Feuchtigkeitsgehalt und Druckfestigkeit von Fichtenholz aus einem Blitzloch. Holz Roh-Werkst 47:479

Bues C-T, Bäucker E, Richter C (2005) Thermische Eigenschaften des Holzes. In: Skripte zum Masterstudiengang Forstwissenschaften, Lehrstuhl Forstnutzung, Technische Universität Dresden, 102pp

Burckhardt H (1879) zit. in Meyer A (1949) Sprachliche und literarische Bemerkungen zum Problem "Drehwuchs". Mitt Schweiz Anstalt für das forstliche Versuchswesen

Burger H (1941) Der Drehwuchs bei den Holzarten. Drehwuchs bei Fichte und Tanne. Bd XX Heft 1. Mitt Schweiz Anst Versuchsw

Burschel P, Huss J (1997) Grundriss des Waldbaus: Ein Leitfaden für Studium und Praxis. 2th revised and erw edn. Parey, Berlin

Butin H (1996) Krankheiten der Wald- und Parkbäume: Diagnose, Biologie, Bekämpfung. 3th revised and extended edn. Thieme, Stuttgart/New York

Butin H, Nienhaus B, Böhmer B (2010) Farbatlas Gehölzkrankheiten; Ziersträucher. Allee- und Parkbäume, Ulmer/Stuttgart

Carpenter RD, Sondermann DL, Rast ED (1989) Grade defects in hardwood timber and logs. Agriculture handbook No. 678. U.S. Department of Agriculture, Forest Service

cgi Deutschland (2012) cgi.Stadtklima-stuttgart.de/mirror/sonne.exe. 30.10.2012

Code of Practice (2011) Praktijkrichtlijnen voor duurzame hootoogst in Surinam. SBB und Tropenbos, Paramaribo

Comvalius LB (2010) Surinamese timber species, characteristics and utilization, 2nd edn. Paramaribo, Celos

Comvalius B (2012) oral communication information. Centre for Agricultural Research in Suriname. 14.09.2012

Dadswell HE, Wardrop AB (1955) The structure and properties of Tension Wood. Holzforschung 9:97

Delorme A (1973) Über die Reichweite von Jahrringchronologien unter besonderer Berücksichtigung mitteleuropäischer Eichenchronologien. In: PZ 48:133–143

Delorme A (1974) Über das Auftreten von Faserstauchungen in Fichtenholz. Forstarchiv 45(7):121–128

DIN (1997a) Rund- und Schnittholz. Messung der Merkmale. EN 1310. DIN Deutsches Institut für Normung eV, Berlin

DIN (1997b) Rund- und Schnittholz Terminologie. Part 8: Begriffe zu Merkmalen von Rundholz. EN 844–8. DIN Deutsches Institut für Normung eV, Berlin

DIN (1997c) Rund- und Schnittholz Terminologie. Part 9: Begriffe zu Merkmalen von Schnittholz. EN 844–9. DIN Deutsches Institut für Normung eV, Berlin

DIN (1997d) Laub-Rundholz Qualitätssortierung. EN 1316–1; 1316–2; 1316–3. DIN Deutsches Institut für Normung eV, Berlin

DIN (1998a) Qualitätssortierung von Nadel-Rundholz. ENV 1927–1; 1927–2; 1927–3. DIN Deutsches Institut für Normung eV, Berlin

DIN (1998b) Rund- und Schnittholz Verfahren zur Messung der Maße. Part 1: Schnittholz. EN 1309–1. DIN Deutsches Institut für Normung eV, Berlin

DIN (1998c) Rund- und Schnittholz. Verfahren zur Messung der Maße. Part 2: Rundholz. PrEN 1309–2./EN 1310 DIN Deutsches Institut für Normung eV, Berlin

C. Richter, *Wood Characteristics: Description, Causes, Prevention, Impact on Use and Technological Adaptation*, DOI 10.1007/978-3-319-07422-1,

DIN (1999a) Qualitätssortierung von Nadel- Rundholz. ENV 1927–1 bis 1927–4 und Laub-Rundholz Qualitätssortierung. EN 1316–1 bis 1316–3. DIN Deutsches Institut für Normung eV, Berlin

DIN (1999b) Schnittholzsortierung nach dem Aussehen (Visuelle Sortierung) für Laubholz – part 1: Eiche und Buche (enthält Änderung A1: 1999); Deutsche Fassung EN 975–1:1995+ A1:1999–10 DIN Deutsches Institut für Normung eV, Berlin

DIN (2001a) Sortierung von Holz nach der Tragfähigkeit. Part 1: Nadelschnittholz, DIN 4074–1 part 3: Sortiermaschinen für Schnittholz, Anforderungen und Prüfung, DIN 4074–3; part 4: Nachweis der Eignung zur maschinellen Schnittholzsortierung, DIN 4074–4. (Normenentwurf vom Mai 2001 als Ersatz für Ausgabe vom Sept. 1989). DIN Deutsches Institut für Normung eV, Berlin

DIN (2001b) Rund- und Schnittholz Terminologie. Part 12: Zusätzliche Begriffe und allgemeiner Index. EN 844–8. DIN Deutsches Institut für Normung eV, Berlin

DIN (2002a) EN1611-1 Sortierung nach dem Aussehen von Nadelholz. Part 1. DIN Deutsches Institut für Normung eV, Berlin

DIN (2002b) Schnittholzsortierung nach dem Aussehen (visuelle Sortierung) von Nadelholz – part 6: Europäische Fichten, Tannen, Kiefern Douglasie und Lärchen (enthält Änderungen 1/2002; Deutsche Fassung EN 1611–1: 1999+A1: 2002–11. DIN Deutsches Institut für Normung e. V. Berlin

DIN (2003) EN 4074–1, Sortierung von Holz nach der Tragfähigkeit. part 1. DIN Deutsches Institut für Normung eV, Berlin

DIN (2004a) Schnittholzsortierung nach dem Aussehen (visuelle Sortierung) von Laubholz – part 2: Pappel; Deutsche Fassung EN 975–2:2004–09 DIN Deutsches Institut für Normung eV, Berlin

DIN (2004b) EN 33 466 Wegweiser für Wanderwege. DIN Deutsches Institut für Normung eV, Berlin/Mai 2004

DIN (2008a) EN 1927–1 Qualitäts-Sortierung von Nadel-Rundholz – part1: Fichten und Tannen. DIN Deutsches Institut für Normung eV, Berlin. 1.6.2008

DIN (2008b) EN 1927–2 Qualitäts-Sortierung von Nadel-Rundholz – part 2: Kiefern. DIN Deutsches Institut für Normung eV, Berlin 1.6.2008

DIN (2008c) EN 1927–3 Qualitäts-Sortierung von Nadel-Rundholz – part 3: Lärchen und Douglasie. DIN Deutsches Institut für Normung eV, Berlin. 1.6.2008

Douglass AE (1919) Climatic cycles and tree-growth. A study of the annual rings in trees in relation to climate and solar activity. Carnegie Inst of Washington 1:1–127

Dujesiefken D (2001) Die häufigsten Irrtümer im Umgang mit Bäumen in der Baumpflege. In: Jahrb. Baumpfl. Thalacker, Braunschweig, 25pp

Dujesiefken D, Liese W (1990) Einfluss der Verletzungszeit auf die Wundheilung bei Buche (Fagus sylvatica L.). Holz Roh-Werkst 48:95–99

Dujesiefken D, Liese W (2006) Die Wundreaktionen von Bäumen. CODIT heute. In: Jahrb. Baumpfl. Haymarket Media, Braunschweig, p 21–40

Dujesiefken D, Liese W (2011) The CODIT principle. New results about wound reactions of trees. Arborist News 20(2):28–30

Dujesiefken D, Peylo A, Liese W (1991) Einfluss der Verletzungszeit auf die Wundreaktionen verschiedener Laubbäume und der Fichte. Forstw Cbl 110:371–380

Durst J (1955) Taschenbuch der Fehler und Schäden des Holzes. Fachbuchverl, Leipzig

DWD (2007) Hochspannung! – Deutscher Wetterdienst, 4th edn. Offenbach. 8/2007

EMPA (1990) Verdeckte mechanische Schäden bei Sturmholz? Internationaler holzmarkt 23:6

Erteld W, Achterberg W (1954) Narbenbildung, Qualitätsdiagnose und Ausformung bei der Rotbuche. Archiv Forstwes 3(7/8):611pp

Finsterbusch E, Thiele W (1987) Vom Steinbeil zum Sägegatter. Fachbuchverl, Leipzig

Fleckinger A, Steiner H (2000) Der Mann aus dem Eis, 4th edn. Folio Verlag Bozen, Wien/Südtiroler Landesmuseum

Fleischer M (2009) Geschichte der Holzernte in Handarbeit. (Zit. Scheifele M [1996] p 129) Projekte Verl., Halle

Frommhold H (2001) Kommentar zu "Rohholzaushaltung Rohholzverkauf" (HKS Brandenburg), 2th revised edn. Min. f. Landwirtsch., Umweltsch. und Raumordnung des Landes Brandenburg (publ.), Potsdam

Fronius K (1991) Betriebsverluste durch Splitterholz. Holz Zentralbl 112:1735

Funke K, Schuster M, Weihs U, Rust S (2011) Zur zeitlichen Entwicklung kodominanter Vergabelungen (Zwiesel) bei Buche (Fagus silvatica). In: Jahrb. Baumpfl. Haymarket Media, Braunschweig

Furniermagazin (2003) Mahagoni-Pyramiden haben bei Fritz Kohl eine lange Tradition. Furniermagazin p 24–27, DRW-Verlag Weinbrenner, Leinfelden-Echterdingen

Furniermagazin (2004) Best of both worlds – Rotierendes Messern als Alternative zur Vertikaltechnik. Furniermagazin p 58–61, DRW-Verlag Weinbrenner, Leinfelden-Echterdingen

Gäbler H (1962) Allgemeiner Forstschutz. Neumann Verl., Radebeul/Berlin

Gayer K, Fabricius L (1949) Die Forstbenutzung. Hamburg, Berlin

Georgi E (1965) Einige Bemerkungen zur Rindennekrose der Buche. Die sozialist Forstwirtsch 15(3):80–87. Dt. Landwirtschaftsverl, Berlin

Gerard A, Liernur AGM (1927) Hexenbesen, ihre Morphologie. Anatomie und Entstehung. Nijgh & van Ditmar´s Uitgevers-Maatschappij, Rotterdam

GOST (1972) Drevesina Poroki. GOST 2140–71. Gosudarstvenni Komitet Standartov Soveta Ministrov SSSR, Moskva

Gottwald H (1983) Maserwuchs – eine pittoreske Laune der Natur. Holz aktuell 4, Danzer. Special printing

Grammel R (1989) Forstbenutzung. Technologie, Verwertung und Verwendung des Holzes. Parey, Hamburg/Berlin

Greyerz H (1919) Das Hagel-, Ton- oder Mändliholz. Schweiz Z Forstwes 75(5–8):113

Griese F (2004) Das Märchen vom hohen Holzpreis. Holz- Zentralbl 97:1322

Grundner F (1882) Untersuchungen über die Querflächenermittlung der Holzbestände. Berlin

Grüner J, Metzler B (2003) Nectria-Arten an Buchenrinde mit Phloemnekrosen. Poster. Albert-Ludwigs-Universität Freiburg, Forstl. Versuchsanst. Baden-Württemberg, Freiburg/Brsg.

Hansen H (2004) Überprüfung der Festigkeitseigenschaften drehwüchsiger Bäume durch Modellsimulation. Thesis. Lehrstuhl Forstnutzung, Technische Universität Dresden

Harris JM (1989) Spiral grain and wave phenomena in wood formation. Springer, Berlin/Heidelberg/New York

Hartig T (1856) Beiträge zur physiologischen Forstbotanik. Über die Vegetationsperioden der Waldbäume und deren Produkte. Allgem. Forst- u. Jagdztg. p 281–296

Hartig R (1901) Über den Drehwuchs der Kiefer. Forstl.- naturwiss. Z 1895, 8. – Holzuntersuchungen. Berlin

Harzmann LJ (1988) Kurzer Grundriss der allgemeinen Tropenholzkunde. S Hirzel Verlag, Leipzig

Hatzianastassiou N, Matsoukas C, Fotiadi A, Pavlakis KG, Drakakis E, Hatzidimitriou D, Vardavas I (2005) Global distribution of earth's surface shortwave radiation budget. Atmos Chem Phys 5:2847–2867

Helling C (1998) Untersuchungen über durch Schnee und Wind induzierte Holzfehler, speziell Faserstauchungen an Kiefer, Lärche und Omorikafichte. Diploma, Technische Universität Dresden, Lehrstuhl Forstnutzung

Helm A (1982) HKS Kurzanleitung zur Rohholzaufbereitung nach gesetzlichen Handelsklassen. DRW Weinbrenner, Stuttgart

Hennig R, Flechsig J (2000) Türme aus Holz. Praktikumsarb. Inst. f. Baukonstruktionen und Holzbau, Technische Universität Dresden, April 2000

Heß RA, Beck R (1927/1930) Forstschutz. Borgmann W (publ.), Neumann/Neudamm

Heussner K-U (1994) Zum Stand der Dendrochronologie in Mecklenburg-Vorpommern. In: Archäologische Berichte aus Mecklenburg-Vorpommern. Deutsches Archäologisches Institut, Berlin (DAI) p 24–30

HKS (2002a) Die Rohholzsortierung in Deutschland, 5th edn. Richtlinie der EWG. Anhang zur Richtlinie. Euting

HKS (2002b) Die Rohholzsortierung in Deutschland, 5th edn. Richtlinie der EWG. Anhang zur Richtlinie, Zusatzbestimmungen der Bundesländer. Euting

Hoadley RB (1990) Holz als Werkstoff, vol 1. Maier, Ravensburg

Holztechnik Fachkunde (2003) Holztechnik Fachkunde, 19th edn. Verl. Europa-Lehrmittel, Haan-Gruiten, p 119–122

Huber B (1941) Aufbau einer mitteleuropäischen Jahrring-Chronologie. Mitt der Akademie der Deutschen Forstwissenschaft 1:110–125

Kehl H (2013) Vegetationsökologie tropischer & subtropischer Klimate/ LV-TWK (vol 8). Inst. f. Ökologie, Technische Universität Berlin

Kessel H (1990) Festigkeitssortierung von Fichtenkantholz aus Sturmbeständen und Normalbeständen mittels Ultraschall. Forschungsprojekt-Nr.: E-90/28 Fachhochschule Hildesheim. DGfH-Nachrichten 50:23–24

Kirbach C (2008) Metallische Befestigungsmittel im Holz – Auswirkungen auf Holzqualität und technologische Prozesse. Diploma, Inst. f. Forstnutzung und Forsttechnik, Technische Universität Dresden

Kirsche A (1994) Vom Glasmacher zum Reifendreher. publ.: Landesstelle für Erzgebirgische und Vogtländische Volkskultur Schneeberg. Verl. Zum Bunten S, Liestal

Kisser J, Steininger A (1952) Makroskopische und mikroskopische Strukturänderungen bei der Biegebeanspruchung von Holz. Holz Roh- Werkst 10(11):414–421

Knigge W (1958) Das Phänomen der Reaktionsholzbildung und seine Bedeutung für die Holzverwendung. Forstarchiv 29:4–10

Knigge W, Schulz H (1959) Methodische Untersuchungen über die Möglich-keit der Drehwuchsfeststellung in verschiedenen Alterszonen von Laubhölzern. Special printing. Holz Roh- Werkst 17:341–351

Knigge W, Schulz H (1962) Die Holzeigenschaften schleimflussgeschädigter Buchen. Allgem. Forstzeitschr, München. Special printing – 12–1962

Knigge W, Schulz H (1966) Grundriss der Forstbenutzung. Parey, Hamburg/Berlin

Knuchel H (1934) Holzfehler. Reprint by Schäfer, Hannover 1995

Knuchel H (1954) Das Holz. Entstehung und Bau, physikalische und gewerbliche Eigenschaften, Verwendung. Sauerländer, Aarau/ Frankfurt Main

Koch G (1999) Sekundäre Veränderungen im Holz dynamisch beanspruchter Fichten (*Picea abies* [L.] Karst.) aus immissionsbelasteten und windexponierten Hochlagenbeständen. Bundesforschungsanstalt für Forst- und Holzwirtschaft Hamburg

Kollmann F (1942) Über das Gefrieren und den Einfluss tiefer Temperaturen auf die Festigkeit der Hölzer. Mitt Akad dt Forstwiss 2:317–336

König E (1957) Fehler des Holzes. Holz-Zentralbl Verlags-GmbH, Stuttgart

König (2000) Standardchronologie der Jahrringbreite der Holzart Eiche in Sachsen. Arbeitsmittel Dendro-Institut an der Technischen Universität Dresden. (unpublished)

Körner M (1961) Untersuchungen an Haselfichte. Diploma, Inst f. Forstnutzung und Forsttechnik, Technische Universität Dresden

Kotte S (2002) Vergleich der Sortierrichtlinien HKS und CEN- Norm durch eine Liegendsortierung. Paper. Inst. f. Forstnutzung und Forsttechnik, Technische Universität Dresden

Kraft G (1884) Beiträge zur Lehre von den Durchforstungen und Lichtungshieben. Kessinger Pub Co (2009), Whitefish, USA

Krahl-Urban J (1953) Drehwuchs bei Eichen und Buchen. Allgem Forstzeitschr VIII:540–542

Kübler H (1959) Die Ursache von Wachstumsspannungen und die Spannungen quer zur Faserrichtung. Holz Roh- Werkst 1:1–9

Kübler H (1962) Schwinden und Quellen des Holzes durch Kälte. Holz Roh- Werkst 20:364pp

Küthe M (1986) Untersuchungen über die Abhängigkeit des erzielbaren Rundholzwertes von der Ästungshöhe bei wertgeästeten Douglasien. Diploma Inst. f. Forstbenutzung Göttingen, Forstwissenschaftl. Fachber. Georg-August-Univ. Göttingen

Landesforstdirektion T (1997) Hinweise zu Aushaltung, Sortierung. Los- und Preisbildung. publ, Thüringer Landesforstdirektion Erfurt

Landesforstverwaltung Baden-Württemberg (2002) Sortiermerkblätter für Stammholz. In: Die Rohholzsortierung in Deutschland, 5th edn. Euting, 154pp

Lange W (1998a) Exsudate von Bäumen – Erzeugnisse der forstlichen Nebennutzung (1). Die Gummen – eine Gruppe wasserlöslicher oder wasserquellbarer Exsudate. Holz-Zentralbl 22:334

Lange W (1998b) Exsudate von Bäumen – Erzeugnisse der forstlichen Nebennutzung (2). Die Kinos – eine Gruppe gerbstoffhaltiger Exsudate. Holz-Zentralbl 23:343

Lange W (1998c) Exsudate von Bäumen - Erzeugnisse der forstlichen Nebennutzung (3). Die Latices –Exsudate mit polyterpenoiden Coagula. Holz-Zentralbl 31:450

Lilienthal J (2004) Der tropische Wald als Lebensraum. Examensarbeit. www.GRIN.com

Lohmann U (2005) Holzhandbuch, 6th updated edn. DRW Weinbrenner, Leinfelden-Echterdingen

Mahler G (1995) Entwicklung der Güteklassen von Stammholz. Staatswald Baden-Württemberg. Research report. Forstliche Versuchsanstalt Baden-Württemberg. Freiburg/Brsg

Mahler G, Richter C, Groß M, Weber A, Maier T, König J (1997) Verfahren zur konservierenden Lagerung von frischem Rundholz. Technische Universität Dresden//Forstl. Versuchs- u. Forschungsanst. Freiburg//Brsg., Europäisches Patentamt München: PCT/DE97/02966, AT: 19.12.1997, Patenterteilung. 14. 11. 2001

Mattheck C (1997) Design in der Natur: Der Baum als Lehrmeister, 3th revised and extended edn. Freiburg/Brsg., Rombach

Matyssek R, Fromm J, Renneberg H, Roloff A (2010) Biologie der Bäume. Ulmer, Stuttgart

Matzek R (1983) Die Herstellung von Maserfurnieren. Holz aktuell 4, Danzer, special printing

Matzek R (1985) Das größte Flugzeug der Welt: Die Spruce Goose ist aus Holz. Holz aktuell 5, 64–67. Danzer, special printing

Mayer-Wegelin H (1956) Die biologische, technologische und forstliche Bedeutung des Drehwuchses der Waldbäume. Forstarchiv 27(12):265–271

Mette H-J (1965) Kommentar zur TGL 15799 Rohholz, Blatt 3 Rohholz. Furnier- und Klangholz. Forstwirtsch. Inst, Potsdam

Mette H-J (1984) Holzkundliche Grundlagen der Forstnutzung. Deutscher Landwirtschaftsverlag, Berlin

Mette H-J, Boss W (1964) Zur Unrundigkeit von Buchenfurnierholz. Holzindustrie 17:83–85

Mette H-J, Kommert R, Stephan G. (o. Jg.) Studienanleitung für das Lehrgebiet Forstnutzung. Technische Universität Dresden III/21/14

Meyer A (1949) Sprachliche und literarische Bemerkungen zum Problem "Drehwuchs". Mitt Schweiz Anstalt für das forstliche Versuchswesen

Müller H-J (1963) Untersuchungen zur Bemessung der wirtschaftlich tragbaren Wilddichte im Walde nach Standort und Wildschaden. Dissertation, Forstl. Fakultät Eberswalde

Münch E (1937) Entstehungsursachen und Wirkung des Druck- und Zugholzes der Bäume. Silva 25:337, 345

N N (2009a) Misteln deuten auf Schwermetalle hin. Holz- Zentralbl 3:46

N N (2009b) Overall management plan 2008–2032. E-Timberindustry Suriname, Concessie Terreinen 217b, 218b, 219, 221, 244a. Public summary. Paramaribo, May 2009 http://www.dennebosflooring.com/

Niemeier L (2013) Optimierung der Holzverwendung in Surinam unter besonderer Betrachtung der Holzmerkmale der wichtigsten surinamesischen Holzarten. Thesis. Universität Hamburg, Inst. f. Weltforstwirtschaft

Niemz P (1993) Physik des Holzes und der Holzwerkstoffe. DRW Weinbrenner, Leinfelden-Echterdingen

Nienhaus F (1979) Lärchen-Degeneration durch Rickettsien ähnliche Bakterien. Allgem Forstz 34:130–132

Oke TR (1978) Boundary layer climates, 2nd edn. Methuen & Co Ltd, London/New York

Paungger J, Poppe T (2000) Vom richtigen Zeitpunkt. Hugendubel, München, p 175

Polge H (1963) A new method for determining the texture of wood. The densitometric analysis of radiographic films. Annales de l'Ecole Nationale des Eaux et Forest de la Recherches et Experiences XX(4):533–581

Pope DJ, Marcroft JP, Whale LRJ (2005) The effect of global slope of grain on the bending strength of scaffold boards. Holz Roh- Werkst 63:321–326

Prien S (1997) Wildschäden im Wald. Ökologische Grundlagen und integrierte Schutzmaßnahmen. Parey, Berlin

Prien S (2006) Waldrehe sind kein Freiwild. Unsere Jagd 9:8–11

Promis A (2009) Natural small-scale disturbance and below-canopy solar radiation effects on the regeneration patterns in a Nothofagus betuloides forest – a case study from Tierra del Fuego, Chile. Dissertation, Albert-Ludwigs-Univ. Freiburg/Brsg

Radkau J, Schäfer I (1987) Holz, ein Naturstoff in der Technikgeschichte. (Zit. Jägerschmid, K. F. V. (1827/1828): Handbuch für Holztransport und Floßwesen zum Gebrauche für Forstmänner und Holzhändler und für solche, die es werden wollen. Tafel 6, Fig. 10–12, 20–22. Karlsruhe.) Rowohlt, Reinbek

Rahmenvereinbarung Werksvermessung (2005) Erläuterung zur Rahmenvereinbarung für die Werksvermessung von Stammholz des Deutschen Forstwirtschaftsrates eV und des Verbandes der Deutschen Säge- und Holzindustrie eV, version 2005-01-14

Rast ED (1982) Photographic guide of selected external defect indicators and associated internal defects in northern red oak. United States Department of Agriculture, Northeastern Forest Experiment Station, Delaware, Ohio

Richter C (2000) Stehendsortierung wertvoller Einzelstämme – eine Möglichkeit zur Verbesserung des Betriebsergebnisses. GIS Aktuell, Belling

Richter C (2002a) Holzmerkmale: Abholzigkeit. Part 1 der Serie "Äußerlich erkennbare Holzmerkmale". Holz-Zentralbl 44:545

Richter C (2002b) Holzmerkmale: Zwieselbildung. Part 2 der Serie "Äußerlich erkennbare Holzmerkmale". Holz-Zentralbl 95:1103

Richter C (2002c) Holzmerkmale: Ovalität/Exzentrizität. Part 3 der Serie "Äußerlich erkennbare Holzmerkmale". Holz-Zentralbl 121:1430

Richter C (2002d) Holzmerkmale: Frostriss/Frostleiste. Part 4 der Serie "Äußerlich erkennbare Holzmerkmale". Holz-Zentralbl 133:1594

Richter C (2003a) Holzmerkmale (8): Der Drehwuchs. Äußerlich erkennbare Holzmerkmale und ihr Einfluss auf die Güte- und Qualitätsklassen-Sortierung. Holz-Zentralbl 31:1189–1190

Richter C (2003b) Auffällige Wuchsmerkmale an Bäumen. In: Beiträge zur Gehölzkunde. HDR Gartenbild Heinz Hansmann, Rinteln, p 110–125

Richter C (2004a) Holzmerkmale (9): Faserstauchung und Faserbruch. Äußerlich erkennbare Holzmerkmale und ihr Einfluss auf die Güte- und Qualitätsklassen-Sortierung. Holz-Zentralbl 9:114

Richter C (2004b) Holzmerkmale (10): Welliger Faser- und Jahrringverlauf. Äußerlich erkennbare Holzmerkmale und ihr Einfluss auf die Güte- und Qualitätsklassen-Sortierung. Holz-Zentralbl 35:442

Richter C (2005a) Holzmerkmale (11): Hohlkehle. Spannrückigkeit und starker Wurzelanlauf. Äußerlich erkennbare Holzmerkmale und ihr Einfluss auf die Güte- und Qualitätsklassen-Sortierung. Holz-Zentralbl 3:38–39

Richter C (2005b) Holzmerkmale (12): Kropf, Knolle. Hexenbesen. Äußerlich erkennbare Holzmerkmale und ihr Einfluss auf die Güte- und Qualitätssortierung. Holz-Zentralbl 35:209–210

Richter C (2006a) Holzmerkmale (14): Kuriositäten an Bäumen. Äußerlich erkennbare Holzmerkmale und ihr Einfluss auf die Sortierung. Holz-Zentralbl 1:3–4

Richter C (2006b) Holzmerkmale (15): Splitterbefall im Holz. Äußerlich erkennbare Holzmerkmale und ihr Einfluss auf die Sortierung. Holz-Zentralbl 3:83

Richter C (2006c) Holzmerkmale, 4th extended edn. Lehrbrief. Inst. f. Forstnutzung u. Forsttechnik, Technische Universität Dresden

Richter C (2008a) Eine alte Eichentruhe erzählt. Holz- Zentralbl 5:126–127

Richter C (2008b) Drehwuchs im Ingenieurholzbau nutzbar. Holz-Zentralbl 9:235

Richter C (2010) Holzmerkmale. 3, extended Edn. DRW-Verlag Weinbrenner, Leinfelden-Echterdingen

Richter C (2012) Studie zur Qualitätsbeurteilung von Wirtschaftsbaumarten aus Primärwäldern in Suriname. Inst. f. Weltforstwirtsch. Universität Hamburg. 15.10.2012

Richter C, Bues C-T (2003a) Holzmerkmale (6): Die Ästigkeit. Äußerlich erkennbare Holzmerkmale und ihr Einfluss auf die Güte- und Qualitätsklassen-Sortierung. Holz-Zentralbl 31:477

Richter C, Bues C-T (2003b) Holzmerkmale (7): Die Astnarbe. Äußerlich erkennbare Holzmerkmale und ihr Einfluss auf die Güte- und Qualitätsklassen-Sortierung. Holz-Zentralbl 50:742–743

Richter C, Bues C-T (2005) Holzmerkmale (13): Blitzschlag. Blitzrinne und Blitzloch. Äußerlich erkennbare Holzmerkmale und ihr Einfluss auf die Sortierung. Holz-Zentralbl 42:535–536

Richter C, Hennig R (2001) Mantel für Holzkonstruktionen mit einem runden Querschnitt. Offenlegungsschrift DE 1001 07 916 A1, AT: 14.2.01; OT: 22.8.02, Dt. Patent- und Markenamt, München

Richter C, Mahler G (2003) Holzmerkmale (5): Die Krummschaftigkeit. Äußerlich erkennbare Holzmerkmale und ihr Einfluss auf die Güte- und Qualitätsklassen-Sortierung. Holz-Zentralbl 6:122

Richter C, Prien S (2006) Holzmerkmale (17): Verbiss-, Fege-, Schälschäden. Äußerlich erkennbare Holzmerkmale und ihr Einfluss auf die Sortierung. Holz-Zentralbl 47:1392–1393

Richter C, König J, Bues C-T (2004) Baumarchäomorphologische Untersuchungen an einer Eichenholztruhe. Sächsische Heimatblätter, 502:120–126. Verl. Gumnior, Chemnitz

Richter HG, Oelker M (2003 onwards): MacroHOLZdata: descriptions, illustrations, identification and informations retrieval. Version: January 2004. http://delta-intkey.com/

Rosenthal M (2009) Entwicklung eines biologisch inspirierten, dreidimensional verformbaren Furniers aus Druckholz. Dissertation. Lehrstuhl für Holz- und Faserwerkstofftechnik, Technische Universität Dresden

Rosenthal M, Bäucker E (2012) Druckholz – Reaktionsholz der Nadelhölzer. Ausgewählte Eigenschaften und wesentliche Unterschiede zum normalen Holzgewebe. Holz- Zentralbl 43:1104–1107

Rubner K (1910) Das Hungern des Kambiums und das Aussetzen der Jahrringe. Naturw. Zeitschr. für Forst- u. Landw.

Rüegsegger P (1962) Die Haselfichte. Eine Studie über ein bekanntes Tonholz. Das Musikinstrument 4:334–335

Rust S, Roloff A, Sommer U (2011) Untersuchungen zur Eignung des h/d- Verhältnisses zur Beurteilung der Sicherheit von Bäumen. In: Dujesiefken D (publ.) Jahrbuch der Baumpflege. Thalacker, Braunschweig p 91–97

RVR (2014) Rahmenvereinbarung für den Rohholzhandel in Deutschland (RVR). publ.: Dt. Forstwirtschaftsrat eV (DFWR) und

Dt. Holzwirt-schaftsrat eV (DHWR) Koordinator: FVA Baden-Württemberg, Abt. Waldnutzung. Freiburg/Brsg 11.12.2014

Sachsse H (1991) Exotische Nutzhölzer. Parey, Hamburg/Berlin

Scheiber C (1965) Tropenhölzer. Fachbuch Verlag, Leipzig

Schirp M (1967) Untersuchungen über die Dimensionsänderungen des Holzes durch Kälte und die Ursachen der Frostrisse der Waldbäume. Dissertation, Universität Hamburg

Schmelzer K (1977) Zier-, Forst- und Wildgehölze. In: Klinkowski M Pflanzliche Virologie 4:276–405

Schmid A (1998) Drehwuchs an der Baumart Eiche. Diploma, Forstwiss. Fak. der Ludwig-Maximilians-Universität München

Schmidt O, Moreth Ute, Butin H, Kehr R (2001) Bakterien in Radialrissen von Stiel-Eiche. Eur J Forest Res 120:1–6. Springer, Berlin/Heidelberg

Schmieder B (2003) 4. Säge- und Wertholzsubmission der Forstverwaltung des Freistaates Sachsen sowie privater und körperschaftlicher Waldbesitzer. Sächs. Min. Umwelt Landwirtsch. Dresden

Schulz H (1958) Merkmale überwachsener Holzfehler in der Baumrinde (I). Holz- Zentralbl 71:919–920

Schulz H (1961) Über die Zusammenhänge zwischen Baumgestalt und Güte des Schnittholzes bei der Buche. Sauerländer, Frankfurt/M

Schute R (1964) Eisen und Blei im Holz. Forst- und Holzwirt 16:345–347

Schute R (1972a) Fäule als Folge von Beschuss an Buche und Fichte. Forst- und Holzwirt 5:108–113

Schute R (1972b) Was kosten Fremdkörper im Holz? Holz- Zentralbl 19:279

Schütt C, Braun E, Kümmerer A, Thies M, Spiecker H (2005) Einschätzung der Holzqualität an stehenden Bäumen. AFZ-Der Wald 6:283–286

Schweingruber FH (2007) Wood structure and environment. Springer, Berlin/Heidelberg

Schweingruber FH (2012) Der Jahrring. Standort, Methodik, Zeit und Klima in der Dendrochronologie. Reprint v. 1983. Kessel, Remagen-Oberwinter

Schweingruber FH, Isler A (1991) Konstruktionsmappe MASSIVHOLZ. publ.: Stiftung Arbeitskreis Schreinermeister. DRW Weinbrenner, Leinfelden-Echterdingen

Schweingruber FH, Kaennel M (1995) Multilingual glossery of dendrochronology. Definitionen in Englisch, Deutsch, Französisch, Italienisch, Portugiesisch und Russisch. Paul Haupt Publishers, Bern/Stuttgart/Wien

Schweitzer A (1960) Kultur und Ethik. Becksche Reihe, Beck, CH

Schwerdtfeger F (1981) Die Waldkrankheiten, 4th edn. Parey, Hamburg Berlin, zit. Eschrich, p 334

Shigo A (1990) Die neue Baumbiologie. Haymarket Media, Braunschweig

Sinn T (2009) Baumkontrollen. Das Modell des Ingenieurbaumes und der biologische Baum. http://www.baumstatik.de/pages/baumkont_sub//zur_wundholzb.html .17.3.09

Spiegel (2006) Luftfahrt; Weißer Riese. Spiegel 52:134, text message

SSLEF (1997) Messung und Sortierung von Rohholz, 3th unveränd. Edn. Sächs. Staatsmin.für Landwirtsch., Ernährung und Forsten (publ.), Dresden

Stein A (1998) Untersuchungen von Faserstauchungen/Faserbrüchen in Fichtenstämmen und ihre Erkennung im Rund- und Schnittholz. Diploma. Inst. für Forstnutzung und Forsttechnik der Technischen Universität Dresden

Stephan G (2012) Die Gewinnung des Harzes der Kiefer, 3th completely revised edn. Kessel, Remagen-Oberwinter

Stepien E, Gadola C, Lenz O, Schär E, Schmid-Haas P (1998) Die Taxierung der Holzqualität am stehenden Baum. Berichte der Eidgenössischen Forschungsanstalt für Wald, Schnee und Landschaft Nr. 344. Birmensdorf

Steuer W (1990) Vom Baum zum Holz, 2th revised edn. DRW Weinbrenner, Leinfelden-Echterdingen

Strasburger E, Noll F, Schenck H, Schimper AFW (1978) Lehrbuch der Botanik, 31st edn. Fischer, Stuttgart/New York

Taffé W (1954) Gütebewegungen des Fichtenholzes verschiedener Standorte und Ertragsklassen in Rheinland-Pfalz. Wissen (Sieg). by Gustav Nising

TGL (1977a) Fachbereichstandard Rohholz. TGL 15799/02 Min. für Land-, Forst- und Nahrungsgüterwirtschaft, Berlin

TGL (1977b) Fachbereichstandard Rohholz. TGL 15799/03 und/04. Min. für Land-, Forst- und Nahrungsgüterwirtschaft, Berlin

TGL (1977c) Fachbereichstandard Rohholz. TGL 15799/05 Min. für Land-, Forst- und Nahrungsgüterwirtschaft, Berlin

TGL (1988) TGL 15799, Sonderheft "Rohholz", part I. Inst. f. Forstwiss. Eberswalde vom 15.04.1988

Thinley C, Palmer G, Vanclay JK, Henson M (2005) Spiral and interlocking grain in *Eucalyptus dunnii*. Holz Roh- Werkst 63:372–379

Thunell B (1951) Über die Drehwüchsigkeit. Holz Roh- Werkst 9(8):293–297

Trendelenburg R (1940) Über Faserstauchungen in Holz und ihre Überwallung durch den Baum. Holz Roh- Werkst 3:209–221

Trendelenburg R (1941) Über innere Schäden (Faserstauchungen und abnorme Sprödigkeit) an einheimischen und tropischen Hölzern. Z für Weltforstwirtsch 8:93–107

Vogelpohl H (1987a) Metallsuchgeräte und deren Anwendung. Holz-Zentralbl 32:457pp

Vogelpohl H (1987b) Ursachen und Auswirkungen metallischer Fremdkörper im Holz. Holz- Zentralbl 25:336

Volkert E (1940) Das Schilfern der Kiefer. Untersuchungen über den Einfluss des Baumalters auf die Holzeigenschaften. Mitt Forstwirtsch u Forstwissensch 3:173–210

Wagenführ R (1966) Anatomie des Holzes. Fachbuchverl, Leipzig

Wagenführ R (2007) Holzatlas, 6th neu bearb. u. erw. edn. Fachbuchverl. Leipzig im Carl Hanser Verl., München

Wagenführ R, Scheiber C (1989) Holzatlas, 3rd edn. Fachbuchverl, Leipzig

Wakin AT, Polubojarinov OI, Solovjov WA (1969) Albom Porokow Drevesini. Izdatel´stvo "Lesnaja Promischlennost´", Moskva

Wanninger J (1989) Untersuchungen über den Drehwuchs an Fichte, Tanne und Kiefer im Bayrischen Wald. Diploma. Forstwiss. Fak. der Ludwig-Maximilians-Universität München

Weidner U (2000) Ergänzungsvorschlag zu den neuen EU- Normen zur Rundholzsortierung – Leitfaden zur Sortierung von Laub- und Nadelrundholz – Inst. f. Holzforschung München, Oktober 2000

Weidner U, Schumacher P, Ritter U, Rathke K-H, Mahler G, Wegener G (1999) Rundholzsortierung nach Europäischen Normen, part 1: Ergänzungsvorschlag zum EU- Entwurf – Untersuchung der Praxistauglichkeit und Auswirkungen auf die Sortierergebnisse. Holz- Zentralbl 145:1950–1951

Weiß H (2006) Bäume kontrollieren, eingehend untersuchen und statistisch beurteilen. Forst und Holz 61(9):351–361

Willing M (1989) Grundlagen der Rohholzsortierung. Technische Universität Dresden, Wissenschaftsbereich Forstnutzung

Willmann U, Mahler G, Wurster M (2001) Güteansprüche am stehenden Stamm im Rahmen der Bundeswaldinventur II. AFZ-Der Wald 56:1024–1026

Wujciak R (1976) Buchenschleimfluss und Holzqualität. Forstarch 47:71–78

Ziegenhals G (2009) Über den Zusammenhang zwischen Jahrringstruktur und akustischen Eigenschaften von Resonanzholz, part 2: Folge der Jahrringbreiten und elastomechanische Eigenschaften. Holztechnologie 50:2

The Author

Dr. rer. Silv. Christoph Richter was born in Chemnitz, Saxony, Germany, in 1943. After secondary school, he completed a 2-year apprenticeship to become a certified forest worker. Then, after high school, he studied at the Dresden University of Technology, Faculty of Forestry in Tharandt from 1963 to 1966. Exmatriculated for political reasons, he returned to forest work. In 1968, he was able to return to his studies and graduated with a degree in forestry engineering in 1970.

He worked as an operations assistant for a state-owned forestry enterprise in 1971, then as a research assistant for the Faculty of Forest Economics and Forest Management at the Dresden University of Technology where he earned his doctorate with the dissertation: "Studies to determine the optimum intensity of thinning operations for spruce trees."

In 1976, the prevailing political climate led him to change career fields and he worked for 12 years at the Dresden Institute of Wood Technology as a research assistant in the area of information/documentation, as well as in research prognosis.

In 1988, he was given the opportunity to resume his work in forestry science at the Technische Universität Dresden, Chair of Forest Management.

After finishing his professional career in 2007, the author focused on the publication of his book "Wood Characteristics," which is now in its third edition. His travels in the tropics ultimately led to the first English edition, expanded to include the wood characteristics of tropical trees.

The author's accomplishments include 80 publications and 10 trademarks. They relate to his scientific fields of expertise: wood characteristics, round timber sorting, wood technology, materials made from renewable resources and recycled products, heat-insulating material with inorganic binding, alternative forestry products and trades, and "technology of intellectual work (heuristics)."

Index

C. Richter, *Wood Characteristics: Description, Causes, Prevention, Impact on Use and Technological Adaptation*,
DOI 10.1007/978-3-319-07422-1, © Springer International Publishing Switzerland 2015

S

T